TM 9-325

DEPARTMENT OF THE ARMY TECHNICAL MANUAL

105-MM HOWITZER M2A1
TECHNICAL MANUAL

by DEPARTMENT OF THE ARMY • *MAY 1948*

DEPARTMENT OF THE ARMY TECHNICAL MANUAL

TM 9–325

This manual supersedes TM 9–325, 19 August 1944, TB ORD 64, 15 March 1944; and those portions of the following technical bulletins pertaining to the equipment covered herein: TB ORD FE 20, 13 December 1944; TB ORD 76, 12 April 1944; TB ORD 139, 10 August 1944; TB ORD 193, 30 September 1944; TB ORD 196, 2 January 1943; TB ORD 280, 4 April 1945; TB ORD 285, 10 April 1945; TB ORD 295, 25 April 1945; TB ORD 300, 24 May 1945; TB ORD 340, 26 March 1947; TB ORD 347, 3 March 1947.

105-MM HOWITZER M2A1

CARRIAGES M2A1 AND M2A2

AND COMBAT VEHICLE

MOUNTS M4 AND M4A1

DEPARTMENT OF THE ARMY • *MAY 1948*

United States Government Printing Office

Washington : 1948

DEPARTMENT OF THE ARMY
Washington 25, D. C., 7 May 1948

TM 9–325, 105-mm Howitzer M2A1; Carriages M2A1 and M2A2; and Combat Vehicle Mounts M4 and M4A1, is published for the information and guidance of all concerned.

[AG 300.5 (25 Feb 48)]

By order of the Secretary of the Army:

OMAR N. BRADLEY
Chief of Staff, United States Army

Official:

EDWARD W. WITSELL
Major General
The Adjutant General

Distribution:
Army:
Tech Sv (2); Arm & Sv Bd (1); AGF (2); OS Maj Comd (10); Base Comd (2); MDW (3); A (ZI) (18), (Overseas) (3); CHQ (2); D (2); FC (1); Class II Instls 9 (3); USMA (2); Sch (5); Tng Ctr (2); PE (Ord O) (5); One (1) copy to each of the following T/O & E's: 6–25; 6–27; 6–29, 6–325; 6–327; 6–329; 9–7; 9–9; 9–12; 9–57; 9–65; 9–67; 9–76; 9–315; 9–316; 9–318; SPECIAL DISTRIBUTION.
Air Force:
USAF (5).
For explanation of distribution formula see TM 38–405.

CONTENTS

This manual supersedes TM 9–325, 19 August 1944, TB ORD 64, 15 March 1944; and those portions of the following technical bulletins pertaining to the equipment covered herein: TB ORD FE 20, 13 December 1944; TB ORD 76, 12 April 1944; TB ORD 139, 10 August 1944; TB ORD 193, 30 September 1944; TB ORD 196, 2 January 1943; TB ORD 280, 4 April 1945; TB ORD 285, 10 April 1945; TB ORD 295, 25 April 1945; TB ORD 300, 24 May 1945; TB ORD 340, 26 March 1947; TB ORD 347, 3 March 1947.

PART ONE

INTRODUCTION

Section I. GENERAL

1. Scope

a. This manual is published for the information of the using arms and services. It contains technical information required for the identification, use, and care of the 105-mm Howitzer M2A1, 105-mm Howitzer Carriages M2A1 and M2A2, Combat Vehicle Mounts M4 and M4A1, ammunition, and accessory equipment.

b. In all cases where the nature of repair, modification, or adjustment is beyond the scope or facilities of the unit, the responsible ordnance service must be informed so that trained personnel with suitable tools and equipment may be provided, or proper instructions issued.

2. Records

a. ARTILLERY GUN BOOK. (1) The Artillery Gun Book (OO Form 5825) is used to keep an accurate record of the matériel. The gun book is stored in gun book cover M539. The book is divided as follows: Record of assignment; battery commander's daily gun record; and the inspector's record of examination.

Note. Record of assignment data must be removed and destroyed prior to entering combat.

These records are important for the following reasons:

(*a*) They inform unit commanders of the condition and service-ability of the weapons under their jurisdiction.

(*b*) They serve as the record of use and maintenance of the matériel and expedite effective maintenance.

(*c*) They serve as a source of technical data to the Ordnance Department for the improvement of weapons, and furnish valuable design data for the development of new weapons.

(2) Complete instructions on how to make entries in the Artillery Gun Book are contained therein. *It is absolutely essential that the gun book be kept complete and up to date, and that the gun book accompany the matériel at all times regardless of where it may be sent.* In order to facilitate proper maintenance of the howitzer and its related matériel (that is, carriage, recoil mechanism, and associate fire control equipment), and to avoid unnecessary duplication of repairs and maintenance, the following additional entries in the gun book are prescribed:

(*a*) A record of completed modification work orders. The record will show the date completed and bear the initial of the officer or mechanic responsible for completion of the modification.

(*b*) A record of the seasonal changes of lubricant and recoil oil in sufficient detail to prevent duplication and afford proper identification by the inspector.

(*c*) The estimated accuracy life of the howitzer is approximately the equivalent of firing 20,000 full service rounds. The reference to OFSB 4–1 in instruction number 6 in the gun book should be deleted.

(*d*) When a removable tube is replaced, an entry of the proof firing and bore diameter data shown on the star gage record which accompanies a new tube will be made in the gun book. Proof facilities will complete and forward star gage records with new tubes, attached in the same general manner as prescribed for gun books ((3) below). The tube serial number and all pertinent firing data for the removed tube which are contained in the gun book will be extracted onto a history of cannon report. History of cannon reports will be extracted in letter form and forwarded through technical channels to the Office, Chief of Ordnance, Attention: ORDFM–Artillery Section, Washington 25, D. C.

(3) The following procedure is prescribed to insure that the Artillery Gun Book will always accompany the matériel whenever it is shipped or transferred from one organization to another:

(*a*) During transfer or shipment, the gun book will be kept in a waterproof envelope securely fastened to the matériel with waterproof tape.

(*b*) Under one of the wrappings of tape, one end of a small tab will be inserted reading "Gun Book Here."

(4) Instructions for making gun book entries and the procedure for keeping the gun book with the cannon whenever it is shipped or transferred from one organization to another must be strictly followed. Ordnance maintenance units, base shops, and depots will insist that the gun book accompany each cannon when it enters their shop for repairs or maintenance.

(5) If a gun book is lost, it will be replaced at once and all available data will be entered in the new gun book. Additional copies of Artillery Gun Book (OO Form 5825, official stock No. 28–F–67990) may be

requisitioned through normal ordnance supply channels. A gun book which has become separated from the weapon to which it pertains and for which efforts to locate the weapon have failed, will be forwarded immediately to Office, Chief of Ordnance, Attention: ORDFM–Artillery Section, Washington 25, D. C.

(6) When the cannon (including breech ring or receiver) is condemned, destroyed, turned in for salvage, or otherwise lost from service, the gun book will be forwarded with proper notation to the Office, Chief of Ordnance, Attention: ORDFM–Artillery Section, Washington 25, D. C. Information contained in the gun book to be returned which pertains to the carriage, recoil mechanism, or other weapon components being retained in service will be extracted and inserted in the gun book pertaining to the replacement cannon.

b. FIELD REPORT OF ACCIDENTS. When an accident involving ammunition occurs during practice, the incident will be reported as prescribed in AR 750–10 by the ordnance officer under whose supervision the ammunition is maintained or issued. Where practicable, reports covering malfunctions of ammunition in combat will be made to the Office, Chief of Ordnance, giving the type of malfunction, type of ammunition, the lot number of the complete rounds or separate loading components, and conditions under which fired.

c. UNSATISFACTORY EQUIPMENT REPORT. Suggestions for improvement in design, maintenance, safety, and efficiency of operation prompted by chronic failure or malfunction of the weapon, spare parts, or equipment, should be reported on WD AGO Form No. 468, Unsatisfactory Equipment Report, with all pertinent information necessary to initiate corrective action. The report should be forwarded to the Office, Chief of Ordnance, Field Service Division, Maintenance Branch, through command channels in accordance with instruction number 7 on the form. Such suggestions are encouraged in order that other organizations may benefit. If WD AGO Form No. 468 is not available, refer to TM 38–650 for list of data required on unsatisfactory equipment report.

Section II. DESCRIPTION AND DATA

3. General

a. The 105-mm howitzer matériel covered in this manual is a mobile, general-purpose field artillery piece, classified as light field artillery, and is used as a divisional weapon. The howitzer is manually operated, single-loaded, air-cooled, and uses semifixed ammunition. The firing mechanism is a continuous pull (self-cocking) type actuated by pulling a lanyard. The recoil mechanism is of the hydro-pneumatic type,

3

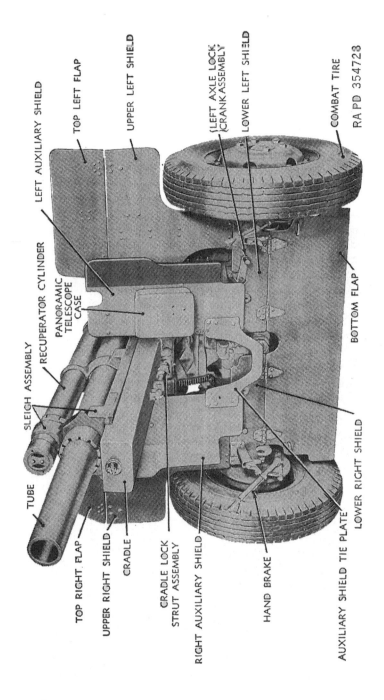

TUBE

SLEIGH ASSEMBLY

RECUPERATOR CYLINDER

PANORAMIC
TELESCOPE
CASE

LEFT AUXILIARY SHIELD

TOP LEFT FLAP

UPPER LEFT SHIELD

(LEFT AXLE LOCK
CRANK ASSEMBLY

LOWER LEFT SHIELD

COMBAT TIRE

RA PD 354723

BOTTOM FLAP

TOP RIGHT FLAP

UPPER RIGHT SHIELD

CRADLE

CRADLE LOCK
STRUT ASSEMBLY

RIGHT AUXILIARY SHIELD

AUXILIARY SHIELD TIE PLATE

LOWER RIGHT SHIELD

HAND BRAKE

Figure 1. 105-mm howitzer M2A1 and carriage M2A2—front view.

4

utilizing a floating piston to separate the oil from the nitrogen. It is used for either direct or indirect fire and can be elevated to high angles to deliver plunging fire on a target.

b. This matériel is composed of the 105-mm howitzer M2A1, with recoil mechanism M2A1, mounted upon either the 105-mm howitzer carriages M2A2 or M2A1 (figs. 1 and 2) or upon the 105-mm howitzer mounts M4 or M4A1 (figs. 3 and 4).

c. When mounted on either the carriage M2A1 (fig. 5) or M2A2 (figs. 6 and 7), it is towed as one unit by a prime mover.

d. When used with either the M4 or M4A1 mount, it is mounted in 105-mm howitzer motor carriage M7 or M7B1 (figs. 8, 9, and 10).

4. Serial Number Information

a. Four serial numbers are required for records concerning the components of this matériel. These components are the howitzer breech ring, the tube, the recoil mechanism, and the carriage or the mount. Serial numbers of all major components will be entered in the gun book.

b. The serial number for the howitzer breech ring is stamped on its top surface (fig. 11).

c. The serial number for the tube is stamped on its chamber end located in the breech ring (fig. 12).

d. The serial number for the recoil mechanism is stamped on a brass plate which is fastened to the left side of recoil mechanism sleigh assembly near the front yoke (fig. 13).

e. The serial number for the carriage or the mount is stamped on a brass plate which is fastened to the left elevating arc (figs. 14 and 15).

5. Differences Between Models

a. The howitzer carriages M2A2 and M2A1 differ as follows:

(1) The carriage M2A2 is equipped with a main shield group composed of right and left upper and lower shields, right and left top flaps, a bottom flap, and right and left auxiliary shields (fig. 1); while the carriage M2A1 is equipped with a right and left main shield and top shields (fig. 2).

(2) Either carriage may be equipped with a screw type traversing mechanism (fig. 20) or with a worm and rack type (fig. 21). Some carriages M2A1 were manufactured with the worm and rack type and those carriages M2A2, which were modified from these carriages, are so equipped.

b. The mounts for combat vehicles differ from the field carriages as follows:

(1) The main shields for the mounts are curved to fit inside the front armor of the vehicle (fig. 10).

5

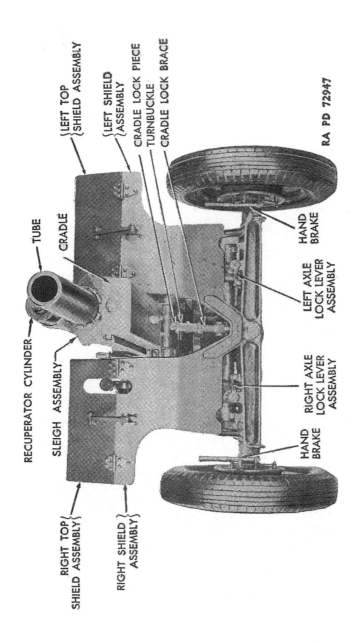

RECUPERATOR CYLINDER

TUBE

SLEIGH ASSEMBLY

CRADLE

{LEFT TOP
}SHIELD ASSEMBLY

{LEFT SHIELD
}ASSEMBLY

CRADLE LOCK PIECE

TURNBUCKLE

CRADLE LOCK BRACE

RA PD 72947

HAND BRAKE

LEFT AXLE LOCK LEVER ASSEMBLY

RIGHT AXLE LOCK LEVER ASSEMBLY

HAND BRAKE

RIGHT TOP SHIELD ASSEMBLY

RIGHT SHIELD ASSEMBLY

Figure 2. 105-mm howitzer M2A1 and carriage M2A1—front view.

6

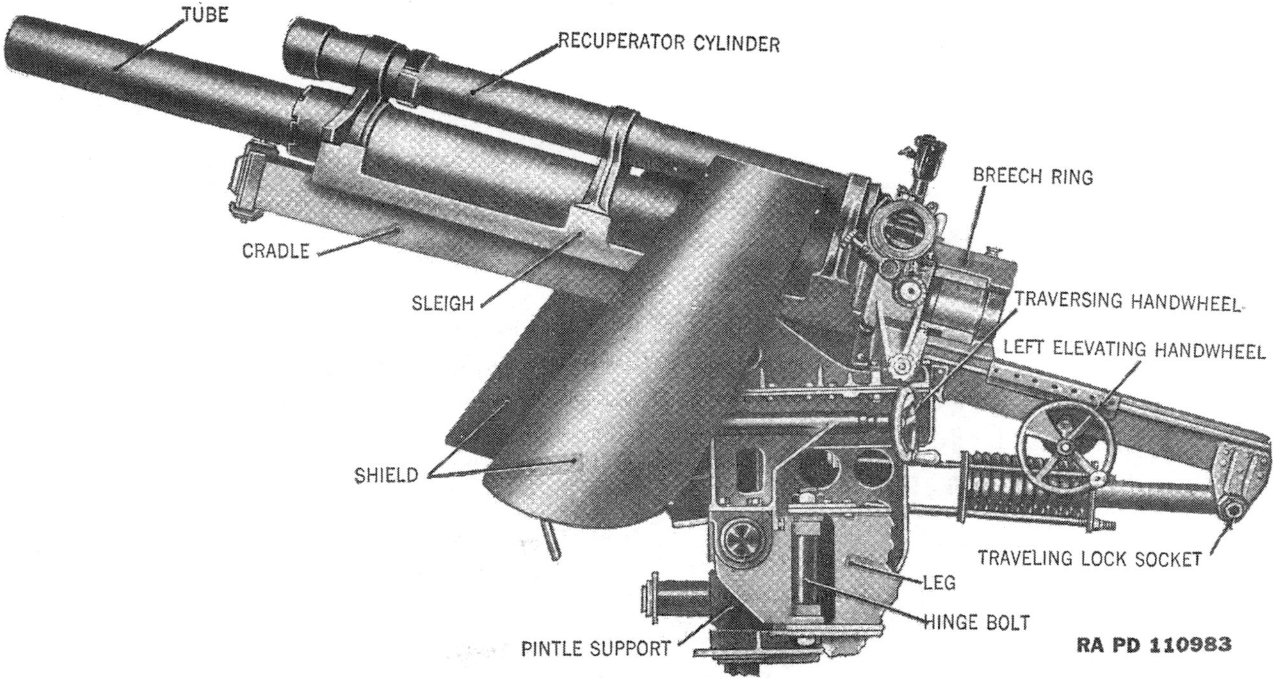

TUBE

RECUPERATOR CYLINDER

CRADLE

SLEIGH

SHIELD

BREECH RING

TRAVERSING HANDWHEEL

LEFT ELEVATING HANDWHEEL

LEG

TRAVELING LOCK SOCKET

HINGE BOLT

PINTLE SUPPORT

RA PD 110983

Figure 3. 105-mm howitzer M2A1 and mount M4—left side view.

7

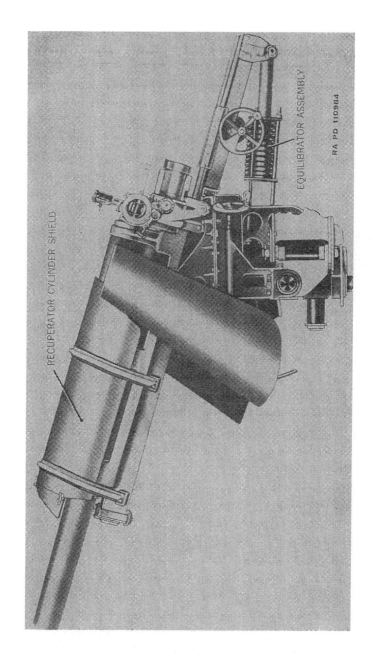

RECUPERATOR CYLINDER SHIELD

EQUILIBRATOR ASSEMBLY

RA PD 110984

Figure 4. 105-mm howitzer M2A1 and mount M4A1—left side view.

8

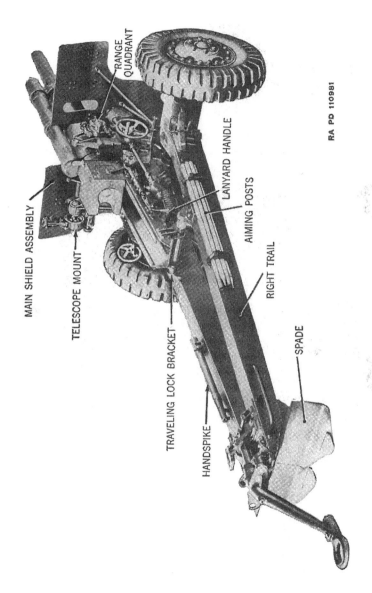

RANGE QUADRANT

MAIN SHIELD ASSEMBLY

TELESCOPE MOUNT

TRAVELING LOCK BRACKET

HANDSPIKE

LANYARD HANDLE

AIMING POSTS

RIGHT TRAIL

SPADE

RA PD 110981

Figure 5. 105-mm howitzer M2A1 and carriage M2A1—right rear view.

10

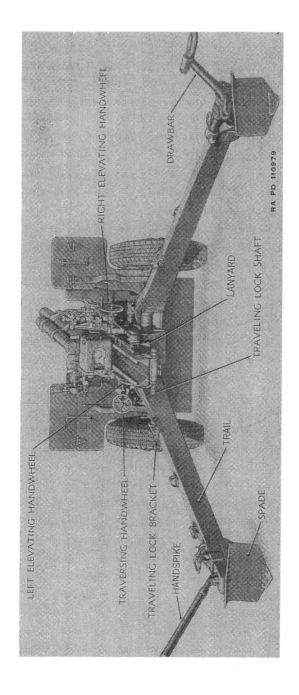

RA PD 110879

Figure 6. 105-mm howitzer M2A1 and carriage M2A2—rear view.

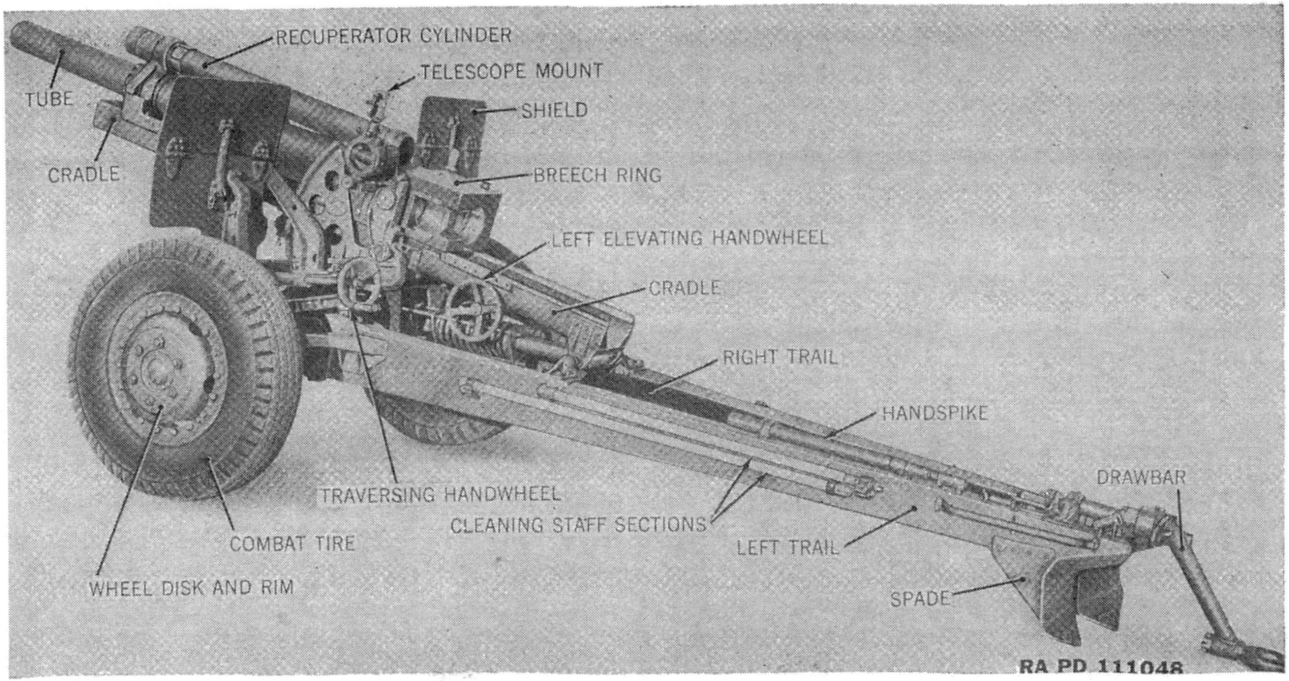

RA PD 111048

Figure 7. 105-mm howitzer M2A1 and carriage M2A2—left side view.

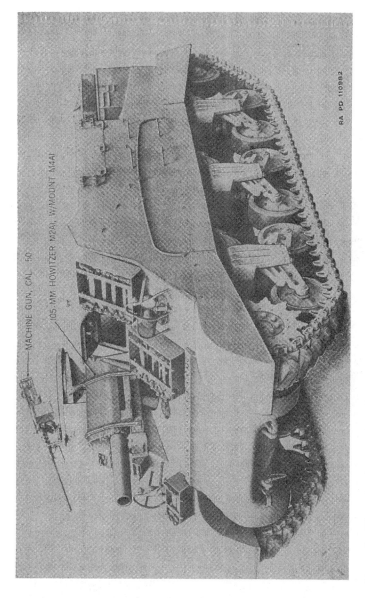

MACHINE GUN, CAL. .50

105-MM HOWITZER M2A1, W/MOUNT M4A1

RA PD 110982

Figure 8. 105-mm howitzer M2A1 and mount M4A1—installed in 105-mm howitzer motor carriage M7—left front view.

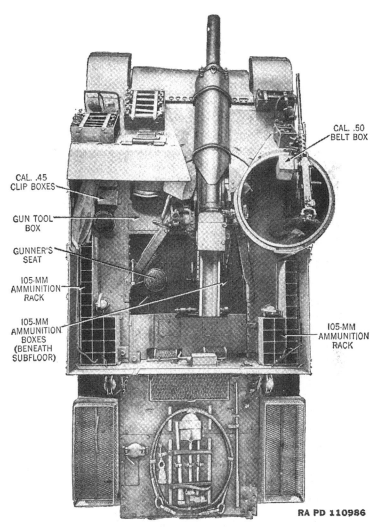

CAL. .50
BELT BOX

CAL. .45
CLIP BOXES

GUN TOOL
BOX

GUNNER'S
SEAT

105-MM
AMMUNITION
RACK

105-MM
AMMUNITION
BOXES
(BENEATH
SUBFLOOR)

105-MM
AMMUNITION
RACK

RA PD 110986

*Figure 9. 105-mm howitzer M2A1 and mount M4A1—installed in
105-mm howitzer motor carriage M7—top view.*

A · RECOIL SLEIGH ASSEMBLY	**K** · CRADLE
B · PANORAMIC TELESCOPE MOUNT	**L** · LEFT ELEVATING HANDWHEEL
C · BREECH RING	**M** · LANYARD
D · BREECHBLOCK OPERATING LEVER ASSEMBLY	**N** · TRAVELING LOCK BRACKET
E · RANGE QUADRANT	**P** · GUNNER'S SEAT
F · FIRING LOCK MI3	**Q** · LEG (CHASSIS PART)
G · RIGHT ELEVATING HANDWHEEL	**R** · TRAVERSING HANDWHEEL
H · NO. I CANNONEER'S SEAT	**S** · HINGE BOLT
J · FIRING MECHANISM	**T** · HINGE BOLT NUT

RA PD 110985

*Figure 10. 105-mm howitzer M2A1 and mount M4—installed in
105-mm howitzer motor carriage M7B1—top rear view.*

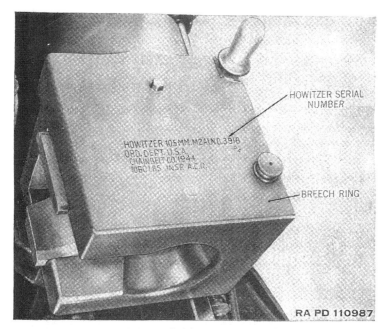

Figure 11. Location of howitzer serial number.

Figure 12. Location of tube serial number.

15

PISTON ROD OUTER NUT

COTTER PIN   SLEIGH

RECOIL MECHANISM
SERIAL NUMBER
RA PD 110988

Figure 13. Location of recoil mechanism serial number.

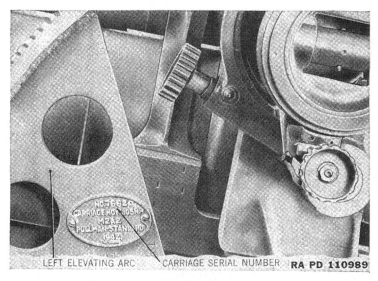

LEFT ELEVATING ARC   CARRIAGE SERIAL NUMBER   RA PD 110989

Figure 14. Location of carriage serial number.

16

Figure 15. Location of mount serial number.

(2) The mounts are supported in the chassis by a heavy pintle support bolted to legs, which are a part of the chassis (fig. 10). The corresponding parts of the field carriages are the equalizing support, axle, wheels, trails, and spades.

(3) The field carriages are equipped with a cradle locking strut to support the cradle (fig. 2) during traveling, while the mounts are not so equipped.

(4) The mounts employ different sighting instruments for direct fire than do the carriages. Indirect fire instruments are the same.

(5) Other slight differences as in handwheel knobs and the type of lanyard cord exist, but they do not affect troop use.

c. The mounts M4A1 and M4 differ as follows:

(1) A shield to protect the recuperator cylinder has been added on the M4A1 (fig. 4).

(2) A heavier equilibrator spring is used on the M4A1 to compensate for the added weight of the recuperator shield.

6. Tabulated Data

a. DATA PERTAINING TO 105-mm HOWITZER M2A1. (1) *General.*

Weight of barrel assembly .... 973 lb
Weight of breech mechanism .... 91 lb
Weight of tipping parts (howitzer, recoil mechanism, cradle,
  sight mount and range quadrant) .... 2,028 lb

17

Length of barrel assembly and breech ring .... 101.35 in.
Length of tube .... 93.05 in.
Diameter of bore .... (4.13 in.) 105-mm
Rifling:
    Length .... (22.5 calibers) 78.02 in.
    Number of grooves 36
    Twist .... Uniform, right-hand, one turn in 20 cals
Type of breechblock .... Horizontal sliding wedge
Type of firing mechanism .... Continuous pull
Ammunition .... For complete ammunition data see section XXII

(2) General performance (average for new howitzer).

Muzzle velocity:
    Shell, H.E. (maximum charge) .... 1,550 ft per sec
    Shell, H.E., A.T. (charge not adjustable) .... 1,250 ft per sec
    (See sec. XXII for data on additional shells.)
Range:
    Shell, H.E. (778.6 mils elevation and zone VII charge) .... 12,205 yds
    Shell, H.E., A.T. (154.6 mils elevation—charge not adjustable) .... 3,500 yds
    (See firing tables and sec. XXII for additional data.)
Estimated accuracy life of tube (equivalent full charge rounds).... 20,000 rounds
Rate of fire (equivalent full charge rounds in cool tube):
    First ½ minute .... 8 rounds per minute
    First 4 minutes .... 4 rounds per minute
    First 10 minutes .... 3 rounds per minute
    Prolonged fire .... 100 rounds per hour

b. DATA PERTAINING TO 105-mm HOWITZER CARRIAGES M2A1 AND M2A2. (1) General.

Weights:
    Howitzer and carriage (complete with accessories) .... (M2A1) 4,475 lb
                                                             (M2A2) 4,980 lb
    Wheel with combat tire (9.00 x 20) .... 287 lb
    At lunette .... 235 lb
Dimensions in traveling position, over-all:
    Length .... 19 ft – 8 in.
    Width .... 7 ft
    Height .... (M2A1)   5 ft
                                                         (M2A2) 5 ft – 2 in.
Road clearance .... (M2A1) 15½ in.
                                                             (M2A2)   14 in.
Turning diameter .... 22 ft
Towed by .... 2½ ton cargo truck or 13-ton high speed tractor, M5
Equilibrator, type .... Spring
Tires:
    Type and size .... 9.00 x 20 combat
    Type and size tubes .... 9.00 x 20 combat
    Air pressure .... 40 lb per sq in.
Brakes, type .... Hand parking

(2) General performance.

Limits of elevation:
    Maximum .... (65 degrees) 1,155 mils
    Depression .... (—5 degrees) —89 mils
Elevation per turn of handwheel .... 10 mils

Limits of traverse:

Right .... (23 degrees) 409 *mils*
Left .... (22½ degrees) 400 mils
Traverse per turn of handwheel .... (Screw type) 19 mils
(Rack type) 21 mils
Diameter of circle of emplacement .... 21 ft

c. DATA PERTAINING TO RECOIL MECHANISM M2A1. (1) *General.*

Type .... Hydro-pneumatic
Grade of recoil oil used .... (See LO 9–325)
Weight (filled) .... 463 lb
Normal length of recoil .... 39 to 42 in.
Maximum allowable recoil .... 44 in.

(2) *Capacities.*

Recoil oil reserve in recuperator .... 1½ fills of (filler) gun

d. DATA PERTAINING TO 105-mm HOWITZER MOUNTS M4 AND M4A1.

(1) *General.*

Dimensions (over-all):
Length:
Howitzer and mount .... 12 ft – 1 in.
Mount M4 .... 9 ft
Mount M4A1 .... 9 ft – 4 in.
Width .... 3 ft – 8 in.
Height:
Mount M4 .... 3 ft – 1 in.
Mount M4A1 .... 3 ft – 5 in.
Equilibrator type .... Spring

(2) *General performance.*

Limits of elevation:
Maximum .... (35 degrees) 622 mils
Depression .... (−5 degrees) −89 mils
Elevation per turn of the handwheel .... 10 mils
Limits of traverse:
Right .... (30 degrees) 534 mils
Left .... (15 degrees) 267 mils
Traverse per turn of the handwheel .... 21 mils

e. LIST OF ON-CARRIAGE SIGHTING AND FIRE CONTROL EQUIPMENT. (See section XXIII for pertinent information.)

MOUNT, telescope, M21A1 and
TELESCOPE, panoramic, M12A2.
QUADRANT, range, M4A1.
MOUNT, telescope, M23 and
TELESCOPE, elbow, M16A1D (for carriage M2A1 or M2A2).
MOUNT, telescope, M42 and
TELESCOPE, elbow, M16A1C (for mount M4 or M4A1).

f. LIST OF OFF-CARRIAGE SIGHTING AND FIRE CONTROL EQUIPMENT NOT COVERED IN THIS MANUAL. (See app. II for pertinent ordnance supply catalogs and technical manuals.)

BINOCULAR (various models).
CHEST, lighting equipment, M21.
CIRCLE, aiming, M1.
COMPASS, M2.

FINDER, range, M7 (with field artillery equipment).

LIGHT, aiming post, M14.

POST, aiming, M1.

QUADRANT, gunner's, M1.

SETTER, fuze, M14 (pertinent information in section XXIII).

SETTER, fuze, M22 (pertinent information in section XXIII).

TABLE, graphical firing, M23.

TELESCOPE, battery commander's, M65.

TELESCOPE, observation, M48 or M49.

THERMOMETER, powder temperature, M1.

WATCH, pocket, 15 or more jewels.

WATCH, wrist, 7 or 15 jewels.

g. DATA PERTAINING TO SUBCALIBER EQUIPMENT. (1) *General*.

Model of gun .... 37-mm subcaliber gun, M13
Weight .... 58 lb
Length of gun .... 33.4 in.
Length of bore .... 29.11 in.
Ammunition .... For complete ammunition data see section XXII

(2) *General performance*.

Muzzle velocity:

Shell, fixed, TP, M92 .... 1,276 ft per sec
Shell, fixed, TP, M63 Mod. 1 .... 1,000 ft per sec

Range:

Shell, fixed, TP, M92 (at 778.6 mils elevation) .... 5,165 yds
Shell, fixed, TP, M63 Mod. 1 (at 778.6 mils elevation) .... 4,980 yds

20

PART TWO

OPERATING INSTRUCTIONS

Section III. GENERAL

7. Scope

Part two contains information for the guidance of the personnel responsible for the operation of this matériel. It contains information on the operation of the equipment with the description and location of the controls and instruments.

Section IV. SERVICE UPON RECEIPT OF MATÉRIEL

8. General

a. Upon receipt of new or used matériel, it is the responsibility of the officer in charge to ascertain whether it is complete and in sound operating condition. A record should be made of any missing parts and of any malfunctions, and any such conditions should be corrected as quickly as possible.

b. Attention should be given to small and minor parts as these are the more likely to become lost or damaged and may seriously affect the proper functioning of the matériel.

c. The matériel should be prepared for service in accordance with the instructions given in paragraph 9 or 10. Any deficiencies should be corrected.

9. New Matériel

a. Howitzers received from storage normally have working parts and unpainted surfaces covered with a corrosion preventive and in addition have the muzzle, breech ring, and equilibrator protected by a tape protective covering.

b. Remove all tape covering from the matériel and disassemble the breech mechanism (par. 77), the firing lock (par. 83), the wheels (par. 99), and the brakes (par. 107). Scrape off as much corrosion preventive as practical from all parts of the matériel and thoroughly clean as outlined in paragraphs 46 and 47. Apply a film of the oil prescribed in paragraph 42 to unpainted surfaces. Pack the wheel bearings (par. 103) and assemble all parts (pars. 78, 84, 100, and 108).

c. Grease and oil in accordance with War Department Lubrication Order 9–325.

d. Perform all detailed inspection and maintenance procedures outlined in paragraphs 53 and 54.

e. Check the gun book for proper entries (par. 2) and check both the gun book and the matériel to see that all prescribed modifications have been performed. See FM 21–6 for list of current modification work orders.

f. Check spare parts, tools, and equipment (sec. X) for completeness and serviceability.

g. Check auxiliary equipment (Part Four) for completeness and serviceability.

10. Used Matériel

a. Perform all the operations listed in paragraph 9.

b. Check all parts of the matériel for signs of excessive wear, damage, missing parts, or corrosion and correct any deficiencies.

c. Pay especial attention to all latches, locks, and catches to see that they are functioning properly.

d. Operate all controls and instruments (pars. 12 and 13) and check carefully for proper functioning.

Section V. CONTROLS AND INSTRUMENTS

11. General

The controls and instruments used to operate the howitzer throughout the cycle of loading, firing, and reloading for continuous firing are located and described in this section. For location and description of sighting controls and instruments used in laying the howitzer, see section XXIII.

12. Controls

a. Breechblock Operating Lever. (1) The L-shaped breechblock operating lever, except for its handle, is inclosed in the top right side of the breech ring when the breech is closed (fig. 16). Its purpose is to open and close the breech.

(2) To open the breech, grasp and depress the operating handle (fig. 16) to release it from the catch, and rotate the operating lever to the rear (fig. 17).

22

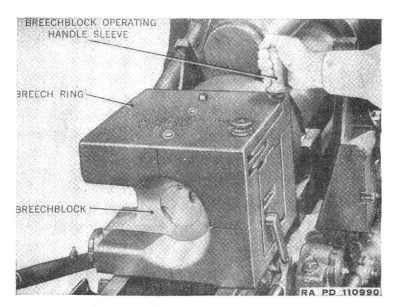

BREECHBLOCK OPERATING
HANDLE SLEEVE

BREECH RING

BREECHBLOCK

RA PD 110990

Figure 16. Breechblock in closed position.

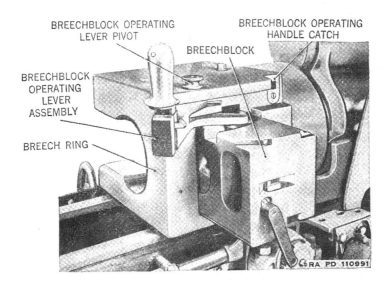

BREECHBLOCK OPERATING
LEVER PIVOT

BREECHBLOCK OPERATING
HANDLE CATCH

BREECHBLOCK

BREECHBLOCK
OPERATING
LEVER
ASSEMBLY

BREECH RING

RA PD 110991

Figure 17. Breechblock in open position.

23

Figure 18. Location of right elevating handwheel.

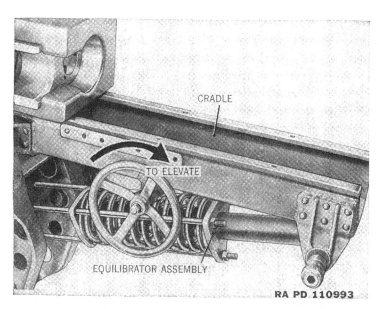

Figure 19. Location of left elevating handwheel.

(3) To close the breech, rotate the operating lever to the front. When closed it automatically engages the catch.

b. ELEVATING HANDWHEELS. (1) The right elevating handwheel is equipped with a knob for easy maneuvering and is located on the right side of the cradle (fig. 18). Its purpose is to elevate or depress the howitzer for changes in elevation.

(2) The left elevating handwheel is located on the left side of the cradle (fig. 19) and is connected with the right elevating handwheel by a system of shafts and bevel gears. The purpose of the left elevating handwheel is to provide the operator with a means for elevating or depressing the howitzer from the side of the weapon having the traversing handwheel.

(3) To elevate the howitzer, turn either the left or the right handwheel in a clockwise direction.

(4) To depress the howitzer, turn either of the handwheels in a counterclockwise direction.

c. TRAVERSING HANDWHEEL. (1) The traversing handwheel is equipped with a knob and is located on the left side of the carriage or the mount (figs. 20 and 21). Its purpose is to traverse the howitzer either to the right or left for changes in azimuth.

(2) To traverse the muzzle to the right, turn the handwheel clockwise if equipped with the screw type traversing mechanism (fig. 20), and counterclockwise if equipped with worm and rack type (fig. 21).

(3) To traverse the muzzle to the left, turn the handwheel in the opposite direction of step (2) above.

d. LANYARD. The lanyard assembly is composed of a cord (or leather strap), an S-hook, and a pear-shaped wooden handle. It is located on the right side of the cradle (fig. 22) and, when pulled, it actuates the firing mechanism to fire the round of ammunition.

e. RESPIRATOR. The respirator is an adjustable air valve located in the rear end of the recoil cylinder (fig. 23). Its purpose is to provide for adjustment of the counterrecoil buffing action to keep the recoiling parts from returning to battery with excessive shock. To adjust the respirator reach under the breech ring, use respirator wrench (41–W–2006–25) to turn the valve counterclockwise to zero (closed) position, then turn clockwise to set to the proper position specified in paragraph 17*e*.

f. BRAKES. Each wheel of the carriage is equipped with a mechanical hand brake of either the clasp type (figs. 1 and 89) or the plunger type (figs. 2 and 88). To set the brake, disengage the catch from the ratchet rack by closing the clasp handle (or depressing the plunger), and pull the brake lever downward to the front until tight. Release the clasp handle (or plunger) to allow the spring-operated catch to engage the ratchet rack to hold the brake in the set position. To release the

Figure 20. Location of screw type traversing handwheel.

Figure 21. Location of rack and worm type traversing handwheel.

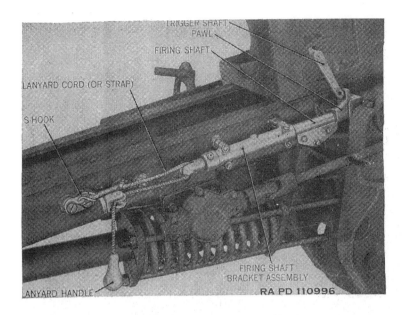

Figure 22. Location of lanyard and firing mechanism.

Figure 23. Location of respirator.

Figure 24. Location of recoil indicator—in marking position.

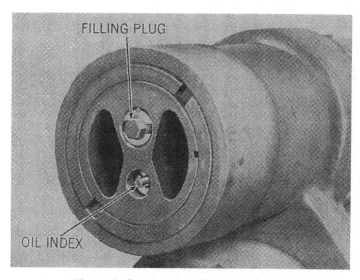

Figure 25. Location of oil index and filling plug.

brake, disengage the catch and push the brake lever upward until the catch engages the square notch in the top end of the ratchet rack.

13. Instruments

a. RECOIL INDICATOR ASSEMBLY. This instrument is used for measuring the length of recoil. It is a spring-loaded plunger seated in a bracket, which is fastened on the right side of the cradle (fig. 24). The plunger can be set to bear against the recoil sleigh assembly so that during recoil it will scrape a mark in the grease on the sleigh rail (par. 21c).

b. OIL INDEX. The purpose of this instrument is to indicate the oil reserve pressure in the recoil mechanism. It has an indicator rod, extending from the recess in the front head of the recuperator cylinder (fig. 25), which is automatically operated by a pinion and rack located in the recuperator cylinder front head (fig. 76). The position of the front end of the indicator rod with reference to the front face of the recuperator cylinder front head indicates the oil reserve as outlined in paragraph 87d.

Section VI. OPERATION UNDER NORMAL CONDITIONS

14. General

Information in this section is concerned with the mechanical steps necessary to prepare the howitzer for firing, fire it, and prepare it for traveling in climates where moderate temperatures and humidity prevail. Preventive maintenance schedules listing all maintenance operations and including operational maintenance prior to and during firing are covered in paragraphs 53 and 54. Operation of subcaliber equipment for training purposes is covered in paragraph 140.

15. Preparation of Howitzer and Carriage for Firing

a. Uncouple the lunette from the pintle of the prime mover (fig. 26). Disengage the drawbar locking shaft (fig. 27) and rotate the drawbar around 180° to the upward position. Reengage the drawbar locking shaft and then lower the trails to the ground.

b. Remove the blackout light-system (fig. 43) from the muzzle end of the tube and from the prime mover and store it in the metal chest provided for small stores.

c. Remove the over-all cover and muzzle cover (fig. 42), after unfastening or unbuckling all straps and opening the zipper fastenings.

d. Disengage the bottom shield latch assemblies (fig. 28) to release the bottom flap.

29

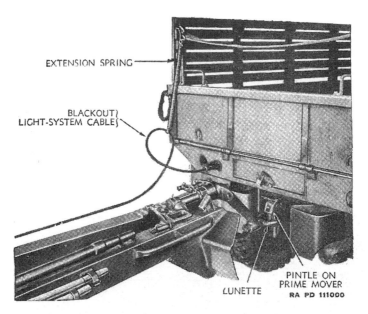

EXTENSION SPRING

BLACKOUT }
LIGHT-SYSTEM CABLE }

LUNETTE

PINTLE ON
PRIME MOVER

RA PD 111000

Figure 26. Field carriage coupled to prime mover.

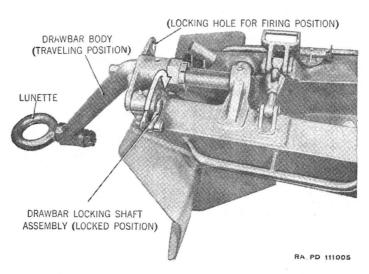

DRAWBAR BODY
(TRAVELING POSITION)

(LOCKING HOLE FOR FIRING POSITION)

LUNETTE

DRAWBAR LOCKING SHAFT
ASSEMBLY (LOCKED POSITION)

RA PD 111005

Figure 27. Disengaging drawbar locking shaft assembly.

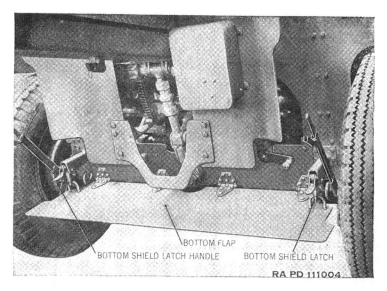

BOTTOM FLAP
BOTTOM SHIELD LATCH HANDLE   BOTTOM SHIELD LATCH
RA PD 111004

Figure 28. Lowering bottom flap to firing position.

e. Disengage the two axle lock crank assemblies, rotate the crank 180° toward the center of the carriage, and reengage the lock in the hole provided (fig. 29), thereby allowing the equalizing axle to function independently of the main axle.

f. Disengage the cradle locking strut from the lower strut latch assembly (fig. 30) and rotate it upward into engagement with the upper strut latch assembly on the under side of the cradle.

DISENGAGE

R. H. AXLE LOCK ASSEMBLY   RA PD 111002

Figure 29. Disengaging axle lock crank assemblies.

31

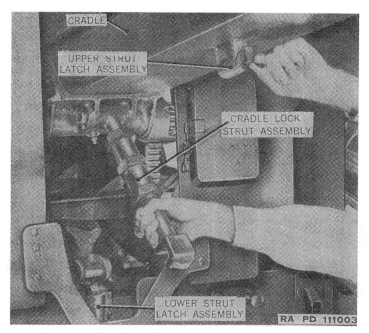

Figure 30. Moving cradle lock strut to firing position.

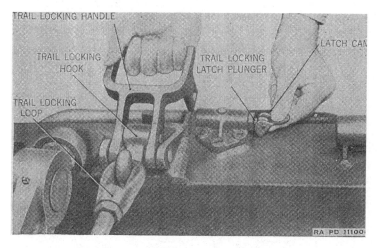

Figure 31. Disengaging trail locking latch.

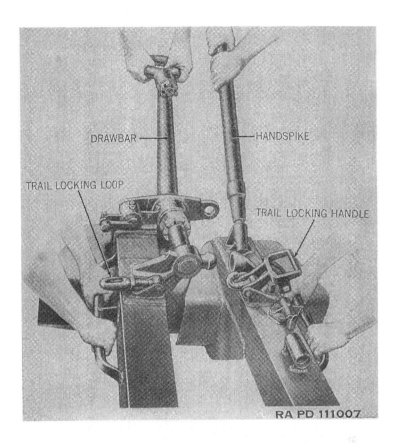

DRAWBAR

HANDSPIKE

TRAIL LOCKING LOOP

TRAIL LOCKING HANDLE

RA PD 111007

Figure 32. Spreading trails to firing position.

g. Disengage the trail locking latch (fig. 31) and unhook the trail locking loop.

h. Insert the handspike in its socket at the rear of the left trail. Using the handspike and the drawbar for leverage, together with the handrails on each trail, lift and spread the trails fully (fig. 32). Lock them in the spread position by inserting the trail locking pins into the forward holes of the trail bumpers, pinning the bumpers to the equalizing support (fig. 33). Holes should be dug for the spades to provide support for the rearward thrust of the howitzer when fired.

i. Lower the top left flap if the aiming (referring) point is obstructed by the flap. On the carriage M2A1, the flap lowers toward the front (fig. 2) and on the carriage M2A2 toward the rear (fig. 34).

j. Install the sighting and fire control equipment as described in section XXIII.

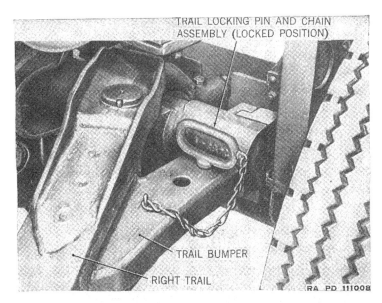

TRAIL LOCKING PIN AND CHAIN
ASSEMBLY (LOCKED POSITION)

TRAIL BUMPER

RIGHT TRAIL

RA PD 111008

Figure 33. Trail locked in firing position by trail locking pin.

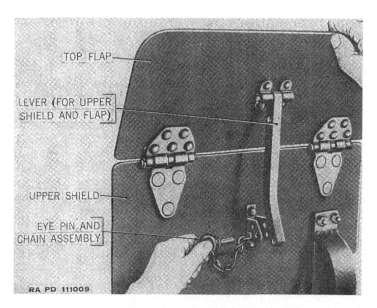

TOP FLAP

LEVER (FOR UPPER
SHIELD AND FLAP)

UPPER SHIELD

EYE PIN AND
CHAIN ASSEMBLY

RA PD 111009

Figure 34. Lowering top left shield flap.

34

k. Tactical instructions for laying the piece are contained in FM 6-75.

l. Set the hand brakes as soon as a sufficient number of rounds have been fired to insure that the spades are firmly seated.

16. Preparation of Howitzer and Mount for Firing

a. If the canvas cover for the forward part of the howitzer is installed, remove it after opening all zipper fastenings and straps.

b. Unbuckle and remove the muzzle cover from the howitzer tube.

c. Disengage the howitzer traveling lock (fig. 35) by turning the traveling lock lever counterclockwise until the socket "B" is clear of the locking shaft ball end. Maneuver the traversing handwheel until socket "A" is clear. Disengage the traveling lock bracket from its position and lay it on the floor out of the way of the crew.

d. Unfasten straps and zipper fastenings of the breech cover and, after the traveling lock is cleared, remove the breech cover.

e. Install the sighting and fire control equipment as described in section XXIII.

f. Tactical instructions for laying the piece are contained in FM 6-74.

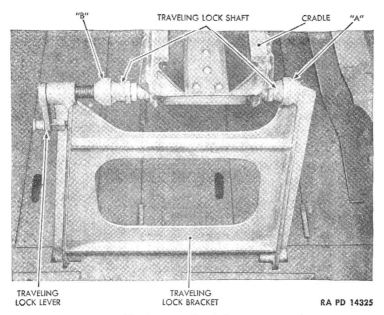

"B"   TRAVELING LOCK SHAFT   CRADLE   "A"

TRAVELING LOCK LEVER   TRAVELING LOCK BRACKET   RA PD 14325

Figure 35. Howitzer traveling lock on motor carriage.

35

17. Inspections and Operations Prior to Firing

a. Wipe the bore and chamber thoroughly dry as outlined in paragraph 47*b*(2).

b. Examine the recoil stuffing box head (fig. 75), the filling plug hole (fig. 25), and the oil index recess (fig. 25) for oil leakage. If there is undue leakage, other than a few drops of dark oil, the howitzer must not be fired. The condition must be reported to ordnance maintenance personnel for correction.

c. Release a small amount of the recoil oil until the index recedes slightly into its recess and then reestablish the correct oil reserve (par. 88), taking care that only enough oil is added to bring the oil index just even with the front face of the recess (par. 87*d*). The howitzer must not be fired with a deficient or excess oil reserve, as to do so may irreparably damage the recoil mechanism.

d. Clean and oil the exposed surfaces of the recoil slides in accordance with LO 9–325.

e. Adjust the respirator (par. 12*e* and fig. 23) using special wrench (41–W–2006–25). The following settings are normally used:

(1) Maximum buffing action is obtained when the respirator is set at "0" position.

(2) Minimum buffing action is obtained when it is set at the "3" position.

(3) For depression or low angle firing, set at "0" or "1" position.

(4) For high angle firing, set at "2" or "3" position.

f. Check sighting and fire control instruments (pars. 127 and 128).

18. Loading the Howitzer

a. Prepare the ammunition for firing as outlined in paragraph 120.

b. Open the breech (par. 12*a*(2)) and insert the round into the chamber as far as it will go. Push with a closed fist to avoid injuring the hand. Close the breech completely (par. 12*a*(3)). Exercise caution with complete rounds to avoid striking the fuze in any manner.

19. Firing the Howitzer

Pull the lanyard once (par. 12*d*). The firing lock is self-cocking and spring-actuated, thereby always delivering the same blow to the primer, regardless of the pull on the lanyard. The firing mechanism returns automatically to the ready position when the lanyard is slackened. In case of failure to fire, refer to paragraph 56.

20. Extracting the Fired Cartridge Case

a. Open the breech smartly. The fired cartridge case will be automatically extracted from the chamber and ejected to the rear of the howitzer.

b. If the extractor fails to extract the fired cartridge case from the chamber when the breech is opened, insert the rammer staff into the muzzle end of the bore and tap the bottom of the inside of the case lightly until it is loosened and can be pushed out of the chamber. Exercise caution when handling the empty case as it may still be hot. See paragraph 57 for correction of failure to extract.

21. Inspections and Operations During Firing

a. Keep all exposed bearing surfaces clean and covered with a thin film of the lubricating oil prescribed in LO 9–325.

b. Check for leakage of recoil oil (par. 17*b*). Constantly verify the complete return of the piece to battery and complete closing of the breechblock. Observe the behavior of the recoil mechanism for length of recoil (*c* below), smoothness of action, and return without shock. See paragraph 61 if howitzer returns to battery with shock, or paragraph 62 if it fails to return to battery. Keep the respirator properly adjusted for counterrecoil buffing action (par. 17*e*).

c. The length of recoil should be between 39 and 42 inches. Recoil of 44 inches or more will permanently damage the recoil mechanism due to physical battering of parts. To measure the length of recoil, grease the under side of the right sleigh rail in line with the recoil indicator, release the spring loaded indicator (fig. 24), fire a round, and measure the length of the mark in the grease. See paragraph 65 if howitzer overrecoils or underrecoils.

22. Unloading the Howitzer

a. A complete round, once loaded, should always be fired in preference to being unloaded, unless military necessity dictates otherwise.

b. A complete round or a stuck projectile will be removed under the direct supervision of an officer, exercising appropriate precautions. Slowly open the breech and catch the complete round, or the cartridge case. If the projectile remains in the tube, fill the chamber with waste, close the breech, and level the howitzer tube. Insert the rammer head of the spring-operated unloading rammer (fig. 40), or of the solid rammer (fig. 39), into the muzzle until it incloses the fuze of the projectile, taking care that it fits properly and has no obstructions. Then push or, if necessary, tap the rammer staff lightly until the projectile is dislodged from its seat. Open the breech and carefully remove the waste and the projectile from the chamber. Remove the rammer and staff from the tube. Segregate the removed round for the inspection of the local ordnance officer to ascertain whether it can be reused.

23. Preparation of Howitzer and Carriages M2A1 or M2A2 for Traveling

a. Remove the telescopes (sec. XXIII) and store them in the panoramic telescope case (fig. 1).

b. Fasten the cleaning staffs to the left trail (fig. 7), the aiming posts, with cover, to the right trail (fig. 5), and check to see that all tools and equipment are properly stored in their chest.

c. Raise the top left flap, if lowered (fig. 34). Raise the bottom flap (fig. 28) until it is engaged by the bottom shield latch assemblies.

d. Withdraw the trail locking pins (fig. 33) and insert them in the rear holes. Release the hand brakes. If removed, insert the handspike in its socket at the rear of the left trail. Using the handspike and drawbar for leverage together with the handrails on each trail, lift and move the trails almost together.

e. Maneuver the elevating handwheel to aline the cradle locking shaft pieces with the cradle traveling lock brackets (fig. 36). See paragraph 95*a* for adjustment, if necessary.

f. Completely close the trails and lock the trail locking latch (fig. 31).

g. Fasten the handspike in its carrying socket on the left trail (fig. 7).

Note. To prevent losing handspike while traveling, a modification may be made by welding a ring (¼ inch stock) around the handspike handle flush and to the rear of the spring lock, 27 inches from the lower end of the handspike. This ring will prevent the handspike handle from slipping through the spring lock and out of the bracket in which the lower end is seated.

h. Disengage the cradle locking strut from the cradle, rotate it downward, and engage it with the lower strut latch assembly (fig. 30). See paragraph 95*b* for adjustment, if necessary.

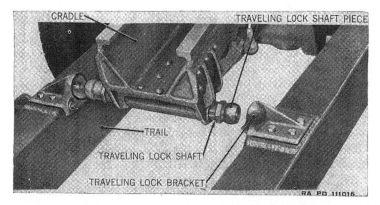

Figure 36. Alining cradle traveling lock and closing trails.

i. Rotate the two axle lock crank assemblies (fig. 29) out toward the wheels and engage the locks in the outer holes. If emplaced on uneven ground it will be necessary to raise one of the trails in order to aline the axle so that the crank assemblies will engage.

j. Install the over-all cover and the muzzle cover (fig. 42).

k. Lift the trails, disengage the drawbar locking shaft (fig. 27), rotate the drawbar around 180 degrees to the downward position, and reengage the drawbar locking shaft. Couple the lunette to the pintle of the prime mover (fig. 26).

l. Install the blackout light-system around the muzzle. Plug the jumper cable into the socket on the prime mover, and attach the extension spring to the prime mover (fig. 26).

24. Preparation of Howitzer and Mounts M4 and M4A1 for Traveling

a. Remove the telescopes (sec. XXIII) and place them in the panoramic telescope case.

b. Fasten the cleaning staffs and the aiming posts, with cover, to the brackets on the compartment rear wall. Check to see that all tools and equipment are properly stored in their chest.

c. Install the breech cover and fasten all straps and zippers.

d. Place the cradle traveling lock bracket in position under the cradle as shown in figure 35. Maneuver the elevating and traversing handwheels to aline the right end of the cradle locking shaft so that it fits snugly into the socket at point "A" (fig. 35). Then turn the traveling lock lever clockwise until the cradle is locked securely in place in the socket at point "B".

e. Install the muzzle cover.

Section VII. OPERATION UNDER UNUSUAL CONDITIONS

25. General

The procedures for the mechanical operation of the howitzer are the same under either usual or unusual climatic conditions. In addition to the normal preventive maintenance service specified throughout this manual, special care in cleaning and lubrication should be observed where extremes of temperature, humidity, and atmospheric conditions are present. Proper cleaning, lubrication, and storage and handling of lubricants not only insure proper operation and functioning but also guard against excessive wear of the working parts and deterioration of the matériel.

26. Operation in Cold Climates

a. In climates where the temperature is consistently below freezing, it is necessary to prepare the howitzer for cold weather operation. Using arms will perform all disassembly prescribed in Part Three, thoroughly clean all the parts, and then change to the greases and lubricants prescribed for cold weather (par. 44a). Ordnance maintenance personnel are responsible for disassembling, cleaning, adjusting, and changing to cold weather lubricants all other assemblies and mechanisms.

b. In cold climates, contamination of lubricants with moisture from snow, rain, or condensation of moisture in partly filled containers is the source of many difficulties. Containers will be kept covered at all times and stored in a warm place, if possible.

c. Matériel placed directly on the ground, ice, or snow, will freeze in place, making it difficult or impossible to move without digging it loose as well as causing serious damage to the tires. Use of the following methods will prevent this condition:

(1) Coat the portion of the spades and aiming posts which contact the ground, ice, or snow, with grease to keep them from freezing in place.

(2) Place a protective layer of waterproof paper, tar paper, roofing paper, straw, hay, or other dry material under the wheels or sleds, tool chests, and accessories, to keep the moisture and ice from coming in contact with them.

(3) The tires will develop a flat surface at the point of contact, necessitating special precautions when starting to travel (g(3) below). To prevent the development of flat surfaces keep the carriage axle blocked up whenever the weapon is inactive.

d. Whenever the metal parts of the matériel or equipment are cold and the surrounding air temperature rapidly becomes warmer or when they are moved into a warmer area, such as a heated building, a condensation of moisture vapor will occur upon the cold surface. This condition is known as "sweating." It can be prevented as follows:

(1) Do not bring any cold matériel indoors unless it is absolutely necessary. It is best to leave it outdoors, but protected from snow with proper covers. Snow-tight lockers at outdoor temperatures are recommended for keeping binoculars, telescopes, and other equipment.

(2) If it is necessary to bring instruments or other equipment from low temperatures to room temperatures, use "anti-condensation" containers. These containers can be specially made tight-fitting, cloth-framed boxes, or any other fairly air-tight containers with heat conducting walls. Place the cold equipment in the container. Have the container at outside temperature so that it will contain cold dry air. Close the top, bring it indoors, and allow it to come to room

temperature. It can be placed near a stove to hasten the warming-up process. The cold dry air expands as it warms, breathing outward, and therefore no warm, humid air from the room comes in contact with the matériel and there is no condensation on it. When the matériel is entirely at room temperature, sweating will not occur when it is removed from the container.

(3) If condensation occurs on cold matériel, it must be disassembled, cleaned, thoroughly dried, and lubricated after it reaches room temperature, to prevent rust or corrosion. Do not operate the matériel before thoroughly drying, as the moisture will form an emulsion with the oil or grease, necessitating removal of the emulsified lubricant and relubricating the matériel. Do not move matériel having moisture caused by condensation on it into the outdoor temperature as the parts will become covered with frost and may not function.

e. Exercise the various controls (sec. V) throughout their entire range, at intervals as required, to aid in keeping the controls from freezing in place and to reduce the effort required to operate them.

f. The preventive maintenance schedules listed in paragraphs 53 and 54 will be followed. In addition, particular attention will be given to the following points:

(1) Protect matériel when not in use with the proper covers, making sure that they are serviceable, in good state of repair (par. 51), and are securely fastened, so that snow or ice will be kept from the operating parts. Provide as much protection as practical for the wheels and tires and other parts of the matériel and associated equipment.

(2) Keep snow and ice from collecting on the matériel, paying particular attention to the teeth of the elevating arcs, pinions, and traversing worm and rack. Cleaning brush M23 is used to clean the gear teeth.

(3) Lubrication of the matériel in cold climates is covered in paragraph 44*a.*

(4) Cleaning of the matériel in cold climates is covered in paragraph 48.

g. In addition to the procedures for traveling outlined in paragraph 23 or 24, particular attention will be given to the following:

(1) Make a thorough inspection and provide as much protection as possible for all parts. See that covers are properly installed and securely fastened.

(2) More than usual care should be taken when traveling over rough terrain because the suspension assemblies will be stiff and may be easily damaged by any unnecessary shock.

(3) Rubber tires are more brittle and easily damaged in cold temperatures. When starting to travel, take care to keep the speed at a minimum until the flat spot in the tire, if present, has been worked out and the tire has regained its natural shape.

(4) Do not fold canvas when wet or frozen. See paragraph 51 for care of canvas.

h. Friction between recoil slides and guides absorbs an appreciable portion of the energy of recoil. Thickened or congealed lubricant increases this friction, shortens recoil, and retards counterrecoil. Be sure that slides and guides are kept clean and are sparingly lubricated.

Caution: Do not remove recoil oil to increase the rate of recoil and counterrecoil.

i. Set the respirator valve at the "3" position (par. 17*e*) when preparing to fire. As the speed of counterrecoil increases due to warming of the recoil oil, readjust the respirator as required.

j. For cold weather operation of sighting and fire control equipment, see paragraph 135.

k. For the use of artillery sleds M3 in cold weather, see paragraph 30.

27. Operation in Hot Climates

a. Constantly observe the operation of the recoil mechanism as prescribed in paragraph 21, to be sure that expansion of the oil due to the heat does not result in an excess reserve, with resultant damage to the mechanism.

b. In addition to the tire maintenance prescribed in paragraph 103, keep tires covered with materials which may be available to protect them from the direct rays of the sun to prevent excessive air pressure and deterioration of rubber.

c. For precautions in handling ammunition in high temperatures, see paragraphs 116 and 117.

28. Operation under Severe Dust or Sand Conditions

a. Excercise particular care to keep sand and dust out of the mechanisms and oil receptacles when carrying out inspection and lubrication operations or when making adjustments and repairs.

b. Keep all covers in place as much of the time as operations permit, making sure that they are in a good state of repair (par. 51) and are securely fastened, to keep as much sand as possible from operating parts.

c. When preparing to fire (par. 17), make sure that all dirt or sand has been cleaned from the bore and chamber; otherwise firing will result in rapid erosion of the bore and early condemnation of the tube.

d. When the howitzer is active, remove all excess lubrication from the elevating arcs and pinions and traversing worm and rack, as there will be less wear from nearly dry operation than from oil contaminated with dust and dirt, because of its abrasive action.

29. Operation under Conditions of High Humidity, Rain and Mud, or Salty Atmosphere

a. Inspect and maintain the matériel frequently as rust and corrosion are greatly accelerated in moist or salty atmospheres. Take care to inspect interiors and small parts of mechanisms.

b. Canvas covers and other items which may deteriorate from mildew or be attacked by insects or vermin will be inspected, cleaned, aired, and dried frequently, as outlined in paragraph 51.

c. Keep ammunition free from mud, corrosion, or foreign matter. Provide proper drainage around the emplacement to keep it as dry as possible.

d. In muddy terrain after traveling, wash off all mud and caked dirt, paying especial attention to all operating parts, latches, locks, and the wheel and brake assemblies. Remove mud from inside the brake drum and brake lever parts (sec. XX) to prevent scoring of drums or malfunctioning of brakes.

e. Install the artillery sleds (par. 30) to aid in moving across very soft or muddy earth. If the prime mover also has insufficient flotation, use appropriate winches to move the piece.

30. Artillery Sleds M3 (T18)

a. GENERAL. Artillery sleds M3 (T18) (fig. 37) are used to facilitate the towing of the howitzer and carriage over swampy areas, soft or boggy ground, and snow and ice. These sleds are issued in pairs, each sled fitting under one wheel. Sleds provide the necessary increased flotation required to prevent the howitzer and carriage from sinking into soft terrain. Although essentially for use over soft or swampy ground, sleds may remain under the carriage wheels for short distances over hard ground. Because the time for installing the artillery sleds is less than five minutes per pair, their removal is advisable when more than a few hundred feet of hard ground are encountered. The

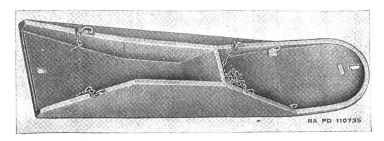

Figure 37. Artillery sleds M3 (T18)

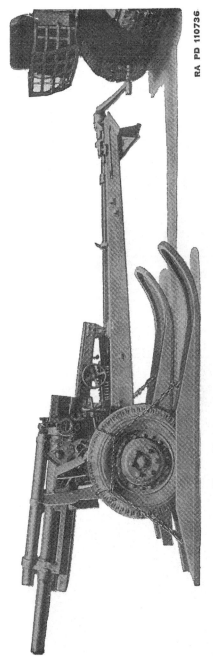

Figure 38. 105-mm howitzer matériel mounted on artillery sled M3.

sleds do not have to be removed for firing. The matériel is towed in the normal manner.

b. DESCRIPTION. Artillery sleds M3 (T18) are made of steel plate and are 10 feet long and 2 feet wide. They have twin keels on a flat bottom surface for steerage and a curved prow to negotiate rough terrain. Wheel wells receive the tires of the howitzer carriage and a harness assembly secures the wheel to the sled (fig. 38).

c. INSTALLATION. Place a sled in front of each carriage wheel, in line with the wheel and pointing in the direction of travel. Tow the carriage onto the sleds so that each wheel fully engages its wheel well. Place the chain harnesses over each wheel and fasten them tightly in place with the shackles (fig. 38).

d. REMOVAL. Remove the chain harnesses from the carriage wheels and back the carriage off the sleds.

31. Preparation for Deep-Water Fording

See TM 9–2853 for instructions to protect matériel when completely submerged in deep-water fording or when engaged in surf landing operations.

Section VIII. DEMOLITION TO PREVENT ENEMY USE

32. General

a. The destruction of the matériel, when subject to capture or abandonment in the combat zone, will be undertaken by the using arm only on authority delegated by the division or higher commander as a command function when such action is deemed necessary as a final resort to keep the matériel from reaching enemy hands.

b. Adequate destruction of artillery matériel means damaging it in such a way that the enemy cannot restore it to usable condition in the combat zone either by repair or by cannibalization. Adequate destruction requires that—

(1) Enough parts essential to the operation of the matériel must be damaged.

(2) Parts must be damaged beyond repair in the combat zone.

(3) The same parts must be destroyed on all matériel, so that the enemy cannot make up one operating unit by assembling parts from several partly destroyed units.

c. The tube and breech are the most vital parts of any piece of artillery. These are the first things to damage. After the tube and breech mechanism in importance comes the recoil mechanism, sighting and fire control equipment, carriage, tires, gun book, and firing tables.

33. Destruction of Barrel Assembly and Recoil Mechanism

a. GENERAL. Four methods, in order of preference, are given for the destruction of the barrel assembly and the recoil mechanism. If possible, the carriage (par. 34) should be destroyed simultaneously. Prior to destruction of matériel, remove all optical sights for evacuation or destruction as prescribed in paragraph 36.

b. METHOD No. 1. (1) Remove filling plug on recoil mechanism and insert the oil release, allowing recoil oil to drain. It is not necessary to wait for the recoil oil to drain completely before firing the weapon as described in step (4) below.

(2) Place the tube at approximately zero degree elevation and insert an armed (safety pin removed) HE antitank grenade M9A1 or an armed antitank rocket M6 into the breech end of the tube with the ogive (nose end) facing back toward the breech, so that the nose is about 21 inches forward of the commencement of rifling; or jam a point detonating HE round in the muzzle.

(3) Load howitzer with HE round, unfuzed if possible. Base-detonating HE shell cannot be used in this method.

(4) Fire the weapon, using a lanyard at least 100 feet long. The person firing should be under cover to the rear of the piece and approximately 20 degrees off the line of fire.

(5) The danger zone is approximately 200 yards and the elapsed time is from 2 to 3 minutes.

c. METHOD No. 2. (1) Insert a charge of 3 to 5 half-pound TNT blocks in the muzzle end of the tube and a charge of 8 to 10 blocks in the chamber, first preparing them for simultaneous detonation by the use of detonating cord passed through the bore to connect the charges.

(2) At the same time, if practicable, also prepare the carriage or mount for simultaneous demolition, as outlined in paragraph 34, by connecting the charge in the muzzle end of the tube and the charge on the top carriage with detonating cord.

(3) Close the breechblock. Plug the muzzle tightly with mud or earth to a distance of approximately 12 inches.

(4) Detonate all charges which have been connected with detonating cord by means of one nonelectric cap inserted in one of the blocks of TNT and igniting the cap with at least 5 feet of safety fuze. (Electric detonation, using electric caps, demolition reel, and exploder, may be used if available.) See FM 5–25 for details of demolition planning and execution.

(5) The danger zone is approximately 200 yards and the elapsed time is from 10 to 15 minutes. Five feet of safety fuze will burn approximately 3½ minutes.

d. METHOD No. 3. (1) Fire one howitzer at the others from a position 200 yards distant. Use HE or HEAT shell. Two or more hits on a vital spot should suffice.

(2) Destroy last howitzer and carriage by best means available.

(3) Enemy salvage of parts is probable with this method.

e. METHOD No. 4. (1) Insert 4 unfuzed incendiary grenades M14 end to end midway in the tube at near zero elevation. Ignite these four grenades by a fifth one equipped with a 15-second delay detonator. The elapsed time is from 2 to 3 minutes. The metal from the grenades will fuze with the tube and fill the grooves.

(2) Destroy the remaining parts by other means.

34. Destruction of Carriage or Mount

a. GENERAL. Whenever possible, destruction should be accomplished simultaneously with the destruction of the barrel assembly and recoil mechanism (par. 33*c*). When this cannot be done, destruction of the barrel assembly and recoil mechanism will have priority.

b. METHOD No. 1. Place 2 unfuzed, boostered, point-detonating HE shells on the top carriage just below the cradle in the vicinity of the pintle. Set the shells upright and place a half-pound TNT block over the booster in each of the shells. Connect the TNT blocks with detonating cord and detonate them with a nonelectric cap ignited by at least 5 feet of safety fuze. The danger zone is approximately 200 yards and the elapsed time is from 3 to 4 minutes. The fuze will burn approximately 3½ minutes.

c. METHOD No. 2. Place 10 half-pound TNT blocks on top carriage just below cradle in vicinity of pintle. Detonate the TNT charge using detonating cord, a tetryl nonelectric cap, and at least 5 feet of safety fuze. The danger zone is approximately 200 yards and the elapsed time is from 4 to 5 minutes. The fuze will burn approximately 3½ minutes.

35. Destruction of Tires

a. GENERAL. Rubber is such a critical item that, whenever matériel is subject to capture or abandonment, an attempt to destroy pneumatic tires must always be made, even if time will not permit destruction of the remainder of the weapon. With adequate planning and training, however, the destruction of tires may be accomplished in conjunction with the destruction of the weapon without increasing the time necessary.

b. METHOD No. 1. (1) Ignite an incendiary grenade M14 under each tire.

(2) To insure the best result when this method is combined with the destruction by TNT of the carriage, be certain that the incendiary fires are well started before detonating the TNT.

c. METHOD No. 2. Damage the tire with an ax, pick, or heavy machine gun fire. Be sure to deflate the tire before damaging with ax or pick. Pour spare gasoline on the tires and ignite.

36. Destruction of Fire Control Equipment

a. All fire control equipment, including optical sights and binoculars, is difficult to replace. It should be the last equipment to be destroyed if there is any chance of personnel being able to evacuate.

b. If evacuation of personnel is made, all possible items of fire control equipment should be carried.

c. If evacuation of personnel is not possible, fire control equipment will be thoroughly destroyed as follows:

(1) Firing tables, trajectory charts, slide rules, and similar items will be thoroughly burned.

(2) All optical equipment will be thoroughly smashed.

37. Destruction of Ammunition

For destruction of ammunition, see TM 9–1901.

PART THREE

MAINTENANCE INSTRUCTIONS

Section IX. GENERAL

38. Scope

Part three contains information for the guidance of the using organizations responsible for the maintenance of this matériel. It contains information needed for the performance of the scheduled inspection, cleaning, lubrication, and preventive maintenance services as well as description of the major groups and assemblies, their authorized disassembly and assembly, and their functions in relation to other components of the equipment.

Section X. ORGANIZATIONAL SPARE PARTS, TOOLS, AND EQUIPMENT

39. Organizational Spare Parts, Tools, and Equipment

a. SPARE PARTS. A set of organizational spare parts is supplied to the using arm for field replacement of those parts most likely to become worn, broken, or otherwise unserviceable.

b. TOOLS AND EQUIPMENT. A set of organizational tools and equipment is supplied to the using arm for maintaining and using the matériel. This set contains items required for disassembly, assembly, cleaning and preserving the 105-mm howitzer matériel. Tools and equipment should not be used for purposes other than prescribed and, when not in use, should be properly stored in the chest and/or roll provided for them.

c. LIST OF SPARE PARTS, TOOLS, AND EQUIPMENT. (1) Spare parts, tools, and equipment supplied for the 105-mm howitzer M2A1 when used with carriages M2A1 or M2A2 are listed in WD Catalog ORD 7 SNL C–21; when used with mounts M4 or M4A1 and mounted on the 105-mm howitzer motor carriage M7, in WD Catalog ORD 7 SNL G–128; and when mounted in the 105-mm howitzer motor carriage M7B1, in WD Catalog ORD 7 SNL G–199. These catalogs are the authorities for requisitioning replacements.

(2) Spare parts and equipment supplied for the on-carriage sighting and fire control equipment are listed in WD Catalogs ORD 7 SNL F–197 for the telescope mounts M21A1 and M23, elbow telescopes M16A1C and M16A1D, and range quadrant, M4A1; ORD 7 SNL F–214 for the panoramic telescope M12A2; and ORD 7 SNL F–256 for the telescope mount M42. These catalogs are the authorities for requisitioning replacements.

40. Specially Designed Tools and Equipment

a. Certain tools and equipment listed in WD Catalogs ORD 7 SNL C–21, ORD 7 SNL G–128, and ORD 7 SNL G–199 are especially designed for maintenance, repair, and general use with the 105-mm howitzer matériel. These tools and equipment are listed in Table I for information only. Table I also lists items of equipment which are especially designed for general use with the on-carriage sighting and fire control equipment. These items of equipment are listed in WD Catalogs ORD 7 SNL F–197, ORD 7 SNL F–214, and ORD 7 SNL F–256. The catalogs referred to will be used for requisitioning replacements.

b. Fuze wrench M7 was replaced by fuze wrench M7A1, and fuze wrench M7A1 is now the substitute for the fuze wrench M18 (T12E1).

(1) All old type fuze wrenches M7 should be salvaged through supply channels. The fuze wrench M7 can be identified by the rectangular shape of the wrench head.

(2) Some of the fuze wrenches M7A1 have been incorrectly marked "M7." All fuze wrenches M7A1 which are marked "M7" should be remarked "M7A1." The M7A1 wrenches can be identified by the curved contour of the wrench head (fig. 39, 41–W–1596–50). The fuze wrenches M7A1 will be salvaged when the fuze wrenches M18 (T12E1) (fig. 39, 41–W–1496–135) are issued.

Table I

| Item | Identifying number | References | | Use* |
		Fig.	Par.	
For Howitzer, 105-mm, M2A1:				
BRUSH, bore, 105-mm, M12.	B168009	39	47b	
COVER, bore brush, M515..	C83757	39	50	
COVER, breech, M212	D41110	....	24	For howitzer on motor carriage.
COVER, muzzle, M4	C6750	....	24	For howitzer on motor carriage.
COVER, muzzle, M333	D7158284	42	23	For howitzer on towed carriage.

Table I (Continued)

Item	Identifying number	References Fig.	References Par.	Use*
RAMMER, cleaning and unloading, M5	C91873	39	22	For howitzer on motor carriage.
RAMMER, unloading, M2...	D39461	40	22	For howitzer on towed carriage.
SIGHT, bore, breech	41–S–3639	39	131	
SIGHT, bore, muzzle	41–S–3645–350	39	131	
STAFF-SECTION, end (61⅝ in. long).	B197241	39	47*b*	Use with bore brush or rammer.
STAFF - SECTION, intermediate (51¹⁵⁄₁₆ in. long)	B197242	39	47*b*	
TARGET, testing, size 39½ x 27½ in.	1–T–283–42	117	132	For howitzer on towed carriage.
TARGET, testing, size 32 x 31 in.	1–T–283–334	118	132	For howitzer on motor carriage.
WRENCH, fuze, M16 (T5).	41–W–1496–115	39	120*b*	
WRENCH, fuze, M18 (T12E1)	41–W–1496–135	39	40*b* and 121*d*	
or				
WRENCH, fuze, M7A1, carb-S. (to be issued in lieu of Fuze Wrench M18 until supply is exhausted)	41–W–1596–50	39	40*b* and 121*d*	
WRENCH, locking ring	41–W–3248–405	41	71	To remove howitzer locking ring.
For Carriage, Howitzer, M2A1 and M2A2:				
COVER, M115A1 (overall).	D7158289	42	23	
HANDSPIKE	5560682	43	15*h*	To spread and shift trails.
LIGHT-SYSTEM, blackout, 6-8v., 24 ft. cable, complete	C93535	43	23	
TOOL, liquid releasing	41–T–3251–611	44	88	To drain reserve oil.
WRENCH, respirator	41–W–2006–25	44	17*e*	To adjust respirator.
For Mount, Howitzer, 105-mm, M4 and M4A1:				
TOOL, liquid releasing	41–T–3251–611	44	88	To drain reserve oil.
WRENCH, respirator	41–W–2006–25	44	17*e*	To adjust respirator.

*Where the use of an item is not indicated the nomenclature is self-explanatory.

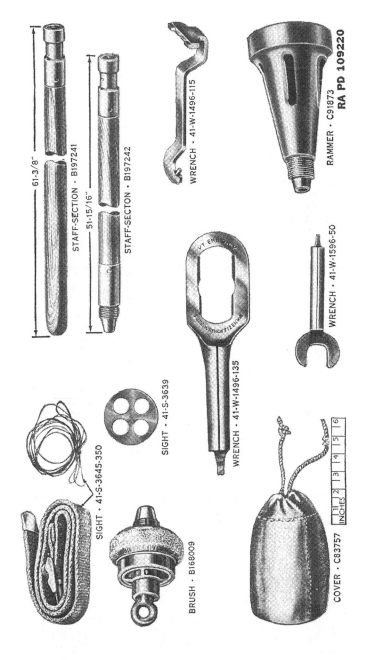

STAFF-SECTION · B197241

STAFF-SECTON · B197242

61-3/8"

51-15/16"

WRENCH · 41-W-1496-115

RAMMER · C91873

RA PD 109220

WRENCH · 41-W-1496-135

WRENCH · 41-W-1596-50

SIGHT · 41-S-3645-350

SIGHT · 41-S-3639

BRUSH · B168009

COVER · C83757

INCHES 1 2 3 4 5 6

Figure 39. Tools and equipment for 105-mm howitzer M2A1.

Table I (Continued)

Item	Identifying number	References Fig.	References Par.	Use*
For Mount, Telescope, M21A1:				
COVER, telescope and mount, M412	D82823	....	133a	
LIGHT, instrument, M19 ...	D43648	108	126c	
For Mount, Telescope, M23:				
LIGHT, instrument, M36 ...	D7690564	113	130a	
For Mount, Telescope, M42:				
LIGHT, instrument, M36 ...	D7690564	113	130a	
For Quadrant, Range, M4A1:				
COVER, telescope and quadrant, M411	D82822	....	129a	
For Telescope, Panoramic, M12A2:				
CHEST, packing, M27	D7691521	....	....	
For Setter, Fuze, M22:				
CASE, Carrying, M55	D7691240	....	....	

*Where the use of an item is not indicated the nomenclature is self-explanatory.

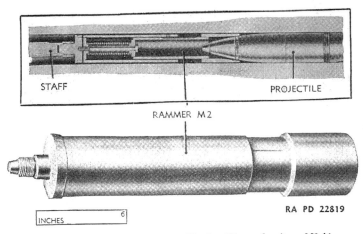

STAFF   PROJECTILE

RAMMER M2

INCHES   6

RA PD 22819

Figure 40. Unloading rammer, M2 for 105-mm howitzer M2A1.

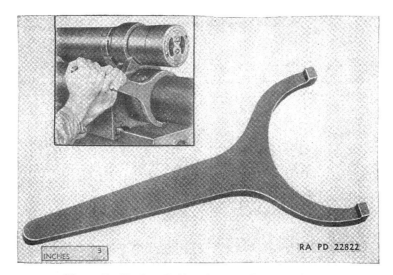

Figure 41. Howitzer locking ring wrench—41-W-3248-405.

Figure 42. 105-mm howitzer matériel with covers installed.

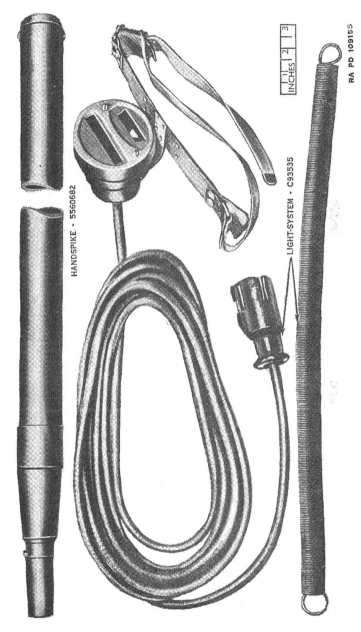

HANDSPIKE - 5560682

LIGHT-SYSTEM - C93535

INCHES 1 2 3

RA PD 109155

Figure 43. Equipment for 105-mm howitzer carriages M2A1 and M2A2.

55

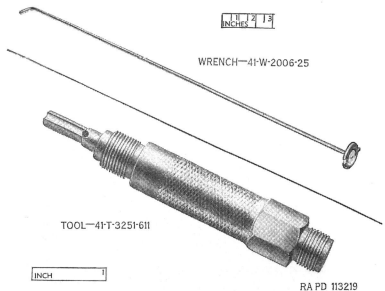

WRENCH—41-W-2006-25

TOOL—41-T-3251-611

RA PD 113219

Figure 44. Tools for 105-mm howitzer carriages M2A1 and M2A2, and mounts M4 and M4A1.

Section XI. LUBRICATION

41. General

This section contains information for properly lubricating and cleaning the matériel. Authorized cleaning and preserving materials are listed in Table II. Prescribed lubricants are listed in the lubrication orders (figs. 46 and 47).

42. Lubrication Orders

a. War Department Lubrication Order 9–325 prescribes organizational lubrication and cleaning maintenance for the howitzer and carriage, and LO 9–749 for the howitzer and combat vehicle mount.

b. The locations of the lubricating fittings are illustrated in figures 48 to 50. They can be identified by circled numbers from 1 to 24 in the illustrations and corresponding numbers added to the lubrication orders, illustrated herein.

c. A lubrication order is placed on, or is issued with, each item of matériel and is to be carried with it at all times. In the event the matériel is received without a lubrication order, one will be requisitioned in conformance with lists in FM 21–6 and instructions in TM 38–405.

43. Lubrication—General Instructions

a. Lubricants are prescribed in the "KEY" on the lubrication order in accordance with three temperature ranges: "above +32° F.," "from +32° F. to 0°F.," and "below 0° F." The time to change grades of lubricants is determined by maintaining a close check on operation of the matériel during the approach to prolonged periods when temperatures will be consistently in the next higher or lower range. Because of the time element involved in preparing for operation at lower prevailing temperatures, a change to lubricants prescribed for a lower range will be undertaken the moment operation becomes sluggish. Ordinarily, it will be necessary to change lubricants only when expected air temperatures will be consistently in the higher or lower range, unless malfunctioning occurs sooner due to lubricants being of improper consistency.

Note. Seasonal changes of lubricants will be recorded in the artillery gun book.

b. Service intervals specified on the lubrication orders and in the maintenance schedules (pars. 53 and 54) are based on normal operating conditions and continuous use of the matériel with frequent firing. Reduce these intervals under extreme conditions, such as excessively high air temperatures, prolonged periods of traveling or firing, operation in sand or dust, immersion in water, or exposure to moisture. Any of these conditions may quickly destroy the protective qualities of the lubricant, and make servicing necessary in order to prevent malfunctioning or damage to the matériel. See section VII for operation under unusual conditions.

c. Clean lubricating equipment both before and after use. Wipe lubricating fittings, oilholes, and surrounding surfaces clean before applying lubricant. Operate the lubricating guns carefully and in such a manner as to insure proper distribution of the lubricant. If lubricating fitting valves stick and prevent the entrance of lubricant, remove the fitting and determine and eliminate the cause. Replace broken or damaged fittings. If a fitting cannot be replaced immediately, cover with tape as a temporary expedient to prevent the entrance of dirt. Lubricating fittings and oilholes are circled with red paint for ready identification. The recoil mechanism filling plug is circled with green paint to indicate use of recoil oil (special). If the plug is circled with yellow paint, notify ordnance maintenance personnel, as yellow paint indicates the use of recoil oil (heavy), which is not prescribed on the lubrication order.

d. Lubricate bearing surfaces equipped with lubricating fittings with the grease gun provided. Lubricate the points equipped with oilholes with the oilcan provided therefor. Lubricate unpainted surfaces by applying the prescribed oil (War Department Lubrication Order 9–325) with a cloth which has been saturated with the oil and then wrung out. Oil the bore by wrapping an oil saturated cloth around the cleaning brush

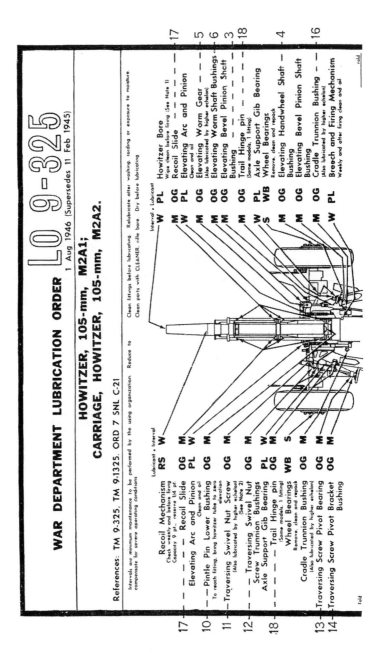

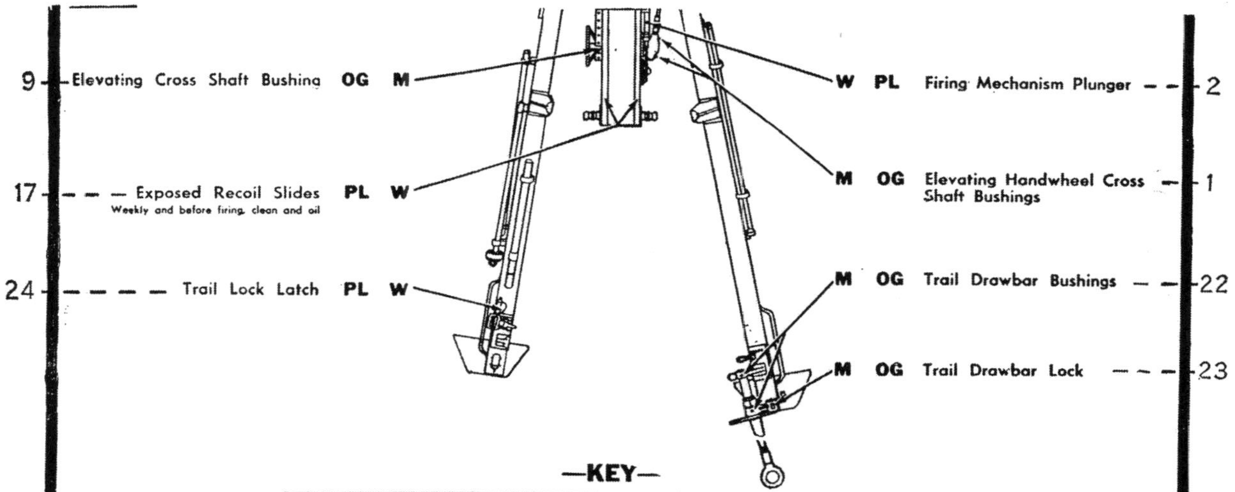

9 — Elevating Cross Shaft Bushing **OG M**

17 — — — Exposed Recoil Slides **PL W**
Weekly and before firing, clean and oil

24 — — — Trail Lock Latch **PL W**

W PL Firing Mechanism Plunger — — 2

M OG Elevating Handwheel Cross — 1
Shaft Bushings

M OG Trail Drawbar Bushings — 22

M OG Trail Drawbar Lock — — 23

RA PD 354729

—KEY—

LUBRICANTS	EXPECTED TEMPERATURES			INTERVALS
	above +32° F.	+32° F. to 0° F.	below 0° F.	
	PL—Medium	PL—Medium	PL—Special	**W**—Weekly
	OG—0	OG—00	OG—00	**M**—Monthly
PL—OIL, lubricating, preservative				**S**—6 Months
OG—GREASE, O. D.				
WB—GREASE, general purpose, No. 2—all temperatures				
RS—OIL, recoil, special—all temperatures				

Figure 45. War Department Lubrication Order.

59

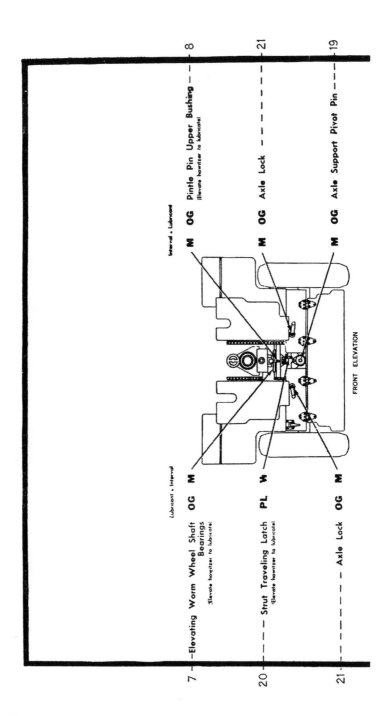

Interval • Lubricant

M OG Pintle Pin Upper Bushing – – 8
(Elevate howitzer to lubricate)

M OG Axle Lock – – – – – 21

M OG Axle Support Pivot Pin – – – 19

FRONT ELEVATION

Lubricant • Interval

OG M –Elevating Worm Wheel Shaft
Bearings
(Elevate howitzer to lubricate)

PL W – – – Strut Traveling Latch
(Elevate howitzer to lubricate)

OG M – – – – – Axle Lock

7

20

21

(1) HOWITZER BORE—After firing and on 3 consecutive days thereafter, clean with CLEANER, rifle bore. After 4th cleaning wipe dry and oil if the howitzer will not be fired within the next 24 hours. Weekly clean with CLEANER, rifle bore, wipe dry and reoil.

(2) TRAVERSING SWIVEL NUT SCREW—Remove plug, if present and insert fitting. Traverse extreme right and lubricate sparingly, then extreme left and lubricate sparingly. Traverse extreme right, clean and apply film of **PL** to exposed traversing handwheel shaft. Do not remove fitting once installed.

OIL CAN POINTS—Weekly, lubricate elevating mechanism universal joints, recoil indicator, trail lock mechanism, traveling lock shaft sockets, traversing and elevating handwheel handles, equilibrator guide rods, hand brake lever assembly, axle lock knob assembly, shield hinges, and locking pins, cradle lock strut hinge pin and strut support atch, shield hinges and latches strut latch pins with **PL**

DISTRIBUTION: AAF (5), AGF (2): T (10): Dept (5); Arm & Sv Bd (1); Tech Sv (2); FC (1), PE (Ord O) (5); Dist 9 (3); Establishments 9 (3) except Am Establishments (0), Gn & Sp Sv Sch (5); A (ZI) (10), (Overseas) (3), CHQ (2); D (2); R 9 (1); Bn 9 (1); C9 (1) except T/O & E 9-17, 9-500-Adm, Dep Am, Distr, Recovery, Bomb Disposal, Am Renovation Plat (GA) (0): One copy to each of the following: T/O & E 6-25; 6-27; 6-29.

For explanation of distribution formula see FM 21-6.

Requisition additional Lubrication Orders in conformance with instructions and lists in FM 21-6.

LUBRICATED AFTER DISASSEMBLY BY HIGHER ECHELON—Cradle trunnion bushings, traversing swivel nut screw, equilibrator spring rod bearing, needle bearings and fulcrum bearing, brake camshaft bearings, elevating worm gear.

Copy of this Lubrication Order will remain with the equipment at all times; instructions contained therein are mandatory and supersedes all conflicting lubrication instructions dated prior to 1 Aug 1946

BY ORDER OF THE SECRETARY OF WAR:

DWIGHT D. EISENHOWER
Chief of Staff

OFFICIAL
EDWARD F WITSELL
Major General
The Adjutant General

RA PD 354730

Figure 46. War Department Lubrication Order.

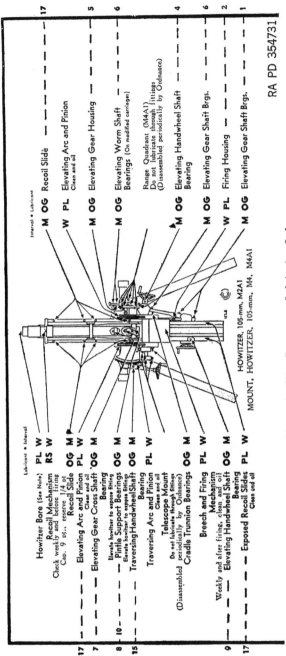

Lubricant ● Interval

Interval ● Lubricant

Howitzer Bore (See Note) — PL W — 17

Recoil Mechanism — RS W
Check weekly and before firing
Cap. 9 pt., reserve 1/4 pt.

Recoil Slide — OG M — M OG — Recoil Slide — 17

Elevating Arc and Pinion — PL W — W PL — Elevating Arc and Pinion — 5
Clean and oil — Clean and oil

Elevating Gear Cross Shaft — OG M — M OG — Elevating Gear Housing — 6
Bearing

Pintle Support Bearings — OG M — M OG — Elevating Worm Shaft — 10
Elevate howitzer to expose fitting — Bearings (On modified carriage)

Traversing Handwheel Shaft — OG M — 15
Bearing — Range Quadrant (M4A1).
Do not lubricate through fittings
(Disassembled periodically by Ordnance)

Traversing Arc and Pinion — PL W — M OG — Elevating Handwheel Shaft — 4
Clean and oil — Bearing

Telescope Mount
Do not lubricate through fittings
(Disassembled periodically by Ordnance)

Cradle Trunnion Bearings — OG M — M OG — Elevating Gear Shaft Brgs. — 6

Breech and Firing — PL W — W PL — Firing Housing — 2
Mechanism
Weekly and after firing, clean and oil

Elevating Handwheel Shaft — OG M — M OG — Elevating Gear Shaft Brgs. — 1
Bearing

Exposed Recoil Slides — PL W — 17
Clean and oil

MOUNT, HOWITZER, 105-mm, M2A1
HOWITZER, 105-mm, M4, M4A1

RA PD 354731

Figure 47. War Department Lubrication Order.

62

RA PD 107875

Figure 48. Lubrication points—numbers 1-10 inclusive.

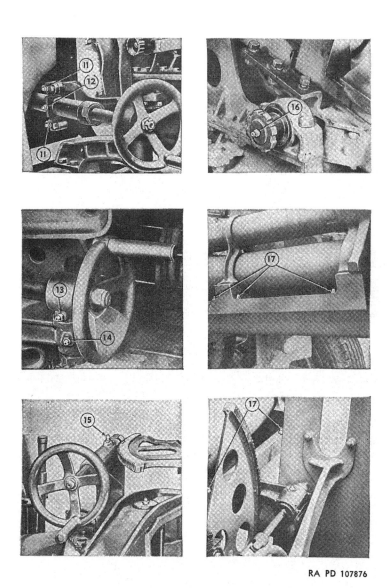

RA PD 107876

Figure 49. Lubrication points—numbers 11–17 inclusive.

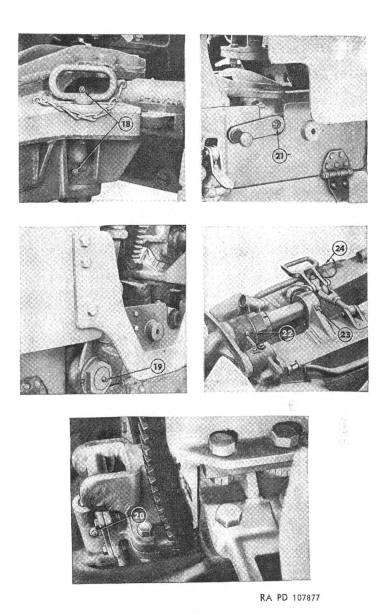

Figure 50. Lubrication points—numbers 18-24 inclusive.

and working it back and forth through the bore. See paragraph 103 for procedure in packing wheel bearings.

e. See WDLO NOTES (fig. 46) for parts which are disassembled and serviced by ordnance maintenance personnel only.

f. Check lubrication point No. 12 (fig. 49) to make sure that a lubricating fitting has been installed. If a plug is installed, remove it and install a lubricating fitting. Traverse the howitzer to the extreme right and lubricate. Traverse the howitzer to the extreme left and again lubricate to insure complete lubrication of the mechanism.

44. Lubrication under Unusual Climatic Conditions

a. Cold Climates. Lubrication at temperatures below 0° F. requires special precautions to prevent malfunction, mechanical failure, or undue wear. Excessive or improper oil or grease on operating parts will thicken or congeal, resulting in sluggish action or complete failure at low temperatures. Use only the lubricants prescribed and apply sparingly. Due to the sparing amount of lubricant used under these conditions, more frequent servicing than is specified for operation under usual conditions (pars. 53 and 54) will be necessary.

b. Hot Climates. Special lubricants will not ordinarily be required at extremely high temperatures, as lubricants prescribed for temperatures above +32° F. provide adequate protection. However, more frequent servicing than specified in paragraphs 53 and 54 is necessary, because the heat tends to dissipate the lubricants.

c. Humid or Salt Air Climates. High humidity, moisture, or salt air tends to contaminate the lubricant, necessitating more frequent servicing than specified in paragraphs 53 and 54.

d. Dusty or Sandy Conditions. Dust and sand when mixed with the lubricant form an abrasive paste and cause extremely rapid wear of moving parts. The amount of lubricant used should be as light and sparing as is practical to obtain proper functioning and the matériel should be serviced more frequently than specified in paragraphs 53 and 54. The elevating arcs and pinions and traversing rack and worm should be wiped nearly dry before operating.

45. Cleaning and Preserving Materials

The cleaners, preservatives, and miscellaneous related items, exclusive of lubricants, required for use with this matériel are listed in Table II. See TM 9–850 for more complete description and use of these materials. This list has been extracted from Department of the Army Supply Catalogs ORD 3, SNL K–1 and ORD 3, SNL K–2, which are used for requisitioning purposes.

Table II

ALCOHOL, ethyl, grade 1 (for telescope eyepieces)	LACQUER, spraying, clear (for steel name plates)
BRUSH, artist's, round, camel's hair, No. 5	NEEDLE, collar, regular bend, 4½ in.
	OIL, neat's foot (for leather goods)
BRUSH, paint, metal bound, flat (medium grade) No. 1 (3 in.)	PAINT, gasoline soluble (various colors as required for camouflage)
BURLAP, jute, 8-oz., 40 in. wide	PALM, sewing
CHALK, railroad, blue, 1 x 4 in. (for marking fire control data on shield)	PAPER, lens tissue, sheets, 7½ x 11 in., 100 sheets per book
CHALK, railroad, white, 1 x 4 in.	SOAP, issue
CLEANER, rifle bore	SOAP, liquid lens cleaning
	SOLVENT, dry-cleaning
CLOTH, abrasive, aluminum oxide, sheets, 9 x 11 in., 5/0–180 (fine) and 3/0–120 (medium)	SPONGE, natural or cellulose
	TAPE, adhesive, nonhygroscopic, O.D.
CLOTH, crocus, sheets, 9 x 11 in.	THINNER, enamel, synthetic (TS)
CLOTH, wiping, cotton	TWINE, jute
ENAMEL, rust-inhibitive, olive drab	WASTE, cotton, white

46. Cleaning—General Instructions

a. GENERAL. (1) See preventive maintenance schedules (pars. 53 and 54) for cleaning intervals.

(2) Rifle bore cleaner is the prescribed cleaner for all surfaces.

(3) During removal from limited storage (par. 2, app. I), dry-cleaning solvent is the preferred cleaner for removing corrosion preventive compound, grease, or oil.

b. PREPARATION OF SOLUTIONS. Rifle bore cleaner and dry-cleaning solvent are used as they come from the container without further preparation.

c. GENERAL PRECAUTIONS IN CLEANING. (1) Rifle bore cleaner will not be diluted.

Note. Rifle bore cleaner is not a lubricant. Parts which require lubrication will be wiped dry and oiled.

(2) Dry-cleaning solvent evaporates quickly, has a drying effect on the skin, and if used without gloves may cause cracks in the skin and in some cases a mild dermatitis. Use only in well-ventilated places. Do not use dry-cleaning solvent to clean parts which have been exposed to powder fouling during firing, because it will not readily dissolve the corrosive salts from the powder and primer compositions.

(3) Avoid getting petroleum products, such as dry-cleaning solvent, engine fuels, or lubricants on rubber parts, as the petroleum will harden, crack, discolor, and/or destroy the rubber. Wash rubber with clean or soapy water.

(4) Never use a solution of lye or other caustic to clean howitzer parts.

(5) The use of gasoline or benzine for cleaning is prohibited.

(6) Gloves should be worn when handling cleaned parts before lubrication, as acid from the hands promotes quick rusting.

(7) Serious damage to sighting and fire control equipment as well as to component parts of the howitzer may result from the use of water, steam, or compressed air from a high pressure hose for cleaning purposes. The safest method of over-all cleaning is by sponging with clean water, or when necessary, with soapy water. Before cleaning the howitzer, remove telescopes and instrument lights from their mounts and install the covers for the mounts and the range quadrant (sec. XXIII).

47. Cleaning—Specific Procedures

a. GENERAL. (1) Refer to general instructions for cleaning (par. 46).

(2) For special precautions when cleaning in temperatures below 32° F., refer to paragraph 48.

b. METHOD OF CLEANING THE BORE AND CHAMBER. (1) Bring the tube to an approximately horizontal elevation and assemble the bore brush to the cleaning staff (fig. 39). Apply the cleaner to the brush, insert it into the bore from the chamber end, and thoroughly scrub all surfaces with a pushing and pulling action. When clean, the bore will have a uniform gray appearance. Do not attempt to obtain a bright polished finish.

(2) To dry the bore and chamber, loosely wrap clean, dry jute burlap around the end of the cleaning staff to approximately the diameter of the bore and sew the burlap securely in place with jute twine. Dry all surfaces of the bore and chamber, replacing the jute burlap as is necessary.

(3) See paragraph 43*d* for procedure in oiling the bore.

c. CLEANING THE BORE AND CHAMBER WITH RIFLE BORE CLEANER. After firing, *when the tube has cooled until it can be touched with the bare hand,* and on 3 consecutive days thereafter, or longer if sweating continues, thoroughly clean the bore and chamber (*b*, above) with rifle bore cleaner. Make sure that all surfaces, including the rifling are left well coated with the cleaner. *Do not wipe dry.* After the fourth cleaning, if the howitzer will probably not be fired within the next 24 hours, wipe dry, and oil. When the weapon is not being fired, thoroughly clean the bore and chamber and renew the oil film weekly.

d. Cleaning the Breech Mechanism and Firing Lock. (1) Disassemble the breech mechanism (par. 77) and the firing lock (par. 83).

(2) After firing, and on 3 consecutive days thereafter, clean the parts with a cloth or sponge saturated in rifle bore cleaner. Use a bath for small parts. Suitable brushes may be used, if available.

(3) Wipe dry and apply a film of the prescribed oil.

(4) Reassemble and install the parts (pars. 78 and 84).

(5) When the howitzer is not being fired, thoroughly clean and renew the oil film weekly.

e. General Cleaning. (1) Wheel bearings and other lubricated surfaces such as recoil slides, elevating arcs, traversing worm and rack, and other greasy surfaces will be cleaned with rifle bore cleaner by applying the cleaner with a saturated cloth or sponge to large parts or as a bath to small parts. Suitable brushes may be used, if available. The parts cleaned will be immediately lubricated as prescribed in the lubrication order.

(2) Painted surfaces in general may be cleaned with clean water or with warm soapy water by swabbing with a saturated cloth or sponge. Rinse parts thoroughly with clean water and wipe dry.

(3) Sighting and fire control instruments will be cleaned as prescribed in paragraph 134.

48. Cleaning under Unusual Climatic Conditions

a. General. The mechanical procedures for cleaning the matériel (par. 47) are the same for all temperatures or climatic conditions; however, certain precautions must be observed when cleaning at temperatures below 32° F.

b. Cleaning with Rifle Bore Cleaner at Temperatures Below 32° F. (1) Do not, under any condition, dilute rifle bore cleaner.

(2) Do not add antifreeze to rifle bore cleaner.

(3) Store the cleaner in a warm place, if practical, and shake well before using.

Section XII. PREVENTIVE MAINTENANCE SERVICE

49. General

Preventive maintenance services prescribed by Army Regulations are a function of using organization echelons of maintenance. This section contains important general preventive maintenance procedures and schedules of preventive maintenance service allocated to crew (1st echelon) and battery (2d echelon) maintenance. Special maintenance of specific groups is covered, when necessary, in the section pertaining to the group.

Special maintenance for operation under unusual climatic conditions is covered in section VII. Battery personnel will disassemble only insofar as is described in this manual. If nature of repair requires further disassembly, it will be accomplished by ordnance maintenance personnel.

50. Common Preventive Maintenance Procedures

The following general preventive maintenance will be observed in addition to that referred to in the schedules in paragraphs 53 and 54:

a. Rust, dirt, grit, gummed oil, and water cause rapid deterioration of internal mechanisms and outer unpainted surfaces. Particular care should be taken to keep all bearing surfaces cleaned and properly lubricated. Remove all traces of rust or corrosion from unpainted bearing surfaces with crocus cloth, which is the coarsest abrasive to be used by the using arm for this purpose.

b. Loose parts will be kept tightened and broken parts replaced or repaired.

c. At least every six months check to see that all modifications have been applied. A list of current modification work orders is published in FM 21–6. No alteration or modification will be made, except as authorized by modification work orders.

d. Check tools, equipment, and spare parts for completeness (sec. X). Replace missing items and turn in for repair all damaged items. Use only tools that are provided and see that they are serviceable. After use, items which are susceptible to rust or corrosion must be thoroughly cleaned as outlined in paragraph 46, coated with a film of the oil prescribed in lubrication order, and stored in their proper chests or tool rolls.

e. Should a shell burst near the howitzer, be sure, before firing the next round, that the weapon has not been damaged. Serious damage will be reported to the ordnance officer.

51. Care of Canvas

To prevent formation of damaging mildew, shake out and air the canvas covers for several hours at frequent intervals. Repair without delay any loose grommets or rips in the canvas. A steel sacking needle and jute twine are furnished for this purpose. Detailed instructions are contained in TM 9–850. Failure to make immediate repairs may allow a minor defect to develop into major damage. Mildewed canvas is best cleaned by scrubbing with a dry brush. If water is necessary to remove dirt, it must not be used until mildew has been removed. If mildew is present, examine fabric carefully by stretching and pulling for evidence of rotting or weakening. If fabric shows weakness it is probably not worth retreatment. If not damaged, retreat the canvas as outlined in TM 9–850.

Oil or grease can be removed by scrubbing with issue soap and warm water. Rinse well with clear water and dry.

Caution. At no time is gasoline or dry-cleaning solvent to be used to remove oil or grease spots from canvas. Wet canvas should be thoroughly dried before folding.

52. Painting

a. Painting is for the purpose of preserving outside nonoperating surfaces of the matériel from which the protective finish has become removed by corrosion, wear, removal of corrosion, or other causes. Painting is also used to prevent reflection of light from parts which are or have become shiny and for disruptive pattern painting for camouflage. Refer to TM 9–2851, Painting Instructions for Field Use, for painting instructions. See FM 5–20D for camouflage painting of artillery matériel. Paints are listed in ORD 3 SNL K–1, which is the authority for requisitioning.

b. When the matériel is to be entirely repainted or spot painted for the prevention of corrosion, use olive drab, synthetic, lustreless enamel and synthetic enamel thinner. Preservative lubricating oil, medium, can be used as a temporary expedient to prevent corrosion of damaged areas until they can be painted.

c. All dirt, oil, or grease should be cleaned from the area to be painted by one of the methods outlined in paragraph 47 and any corrosion should be removed with aluminum oxide abrasive cloth or sandpaper, leaving the area clean and dry. Thin the paint to a painting consistency and carefully brush it on the surface to be painted, taking care to prevent running or dripping onto other parts of the matériel. Do not paint over name plates or serial numbers or polished surfaces which are lubricated in accordance with the lubrication order (par. 42).

d. Name plates which have become corroded or rusty will be carefully cleaned and coated with clear lacquer.

e. Sighting and fire control instruments will not be painted by the using arms.

53. Crew (First Echelon Maintenance) Schedules

The following schedule contains information for inspection and maintenance of points for which the crew is responsible. Detailed instructions are included, or the proper paragraph is indicated. Service intervals are based on active use under usual climatic conditions. Reduce intervals for unusual climatic conditions as outlined in section VII. Extend the intervals when not in use.

Point	Preventive maintenance	Detailed instructions
	Before Firing	
Bore and chamber	Wipe dry	Par. 17.
Breech and firing mechanisms.	Check for proper functioning....	Pars. 74 and 80.
Recoil mechanism	Check for excessive oil leakage. Drain and reestablish oil reserve. Adjust the respirator.	Pars. 17e and 88.
Recoil slides	Clean and oil exposed surfaces ..	Pars. 17 and 42.
	During Firing	
Recoil mechanism	Observe behavior and check for (1) smooth operation, (2) length of recoil, (3) complete return to battery without shock, and (4) excessive oil leakage.	Par. 21.
Recoil slides	Keep exposed surfaces lubricated.	Pars. 21 and 42.
	After Firing	
Bore and chamber	Clean thoroughly daily for 3 days or until sweating ceases. Oil as prescribed.	Pars. 42 and 47.
Breech mechanism	Disassemble, clean, and oil daily for 3 days or until sweating ceases.	Pars. 42, 47, 75, and 83.
Elevating arcs and pinions.	Clean and oil	Pars. 42 and 47.
Howitzer in general	Inspect over-all and correct any deficiencies.	Pars. 50 and 52.
Recoil mechanism	Inspect for all leakage. Turn respirator to closed position.	Pars. 12e, 60, and 92.
Recoil slides	Clean and oil exposed surfaces..	Pars. 42 and 47.
Traversing worm and rack.	Clean and oil	Pars. 42 and 47.
	Weekly Service	
Bore and chamber	Examine for evidence of powder fouling or corrosion. Clean and oil.	Pars. 42, 47, and 73.
Breech mechanism and firing lock.	Examine for evidence of corrosion or other damage. Check for clean smooth operation. Clean and oil.	Pars. 42, 47, 79, and 85.
Carriage or mount	Observe general condition. Check for cleanliness, proper lubrication, and condition of paint.	Pars. 46, 50, and 52.
Covers	Check for proper installation and condition of canvas.	Pars. 23, 24, 51, 130e, and 133.

Point	Preventive maintenance	Detailed instructions
Elevating mechanism ...	Check for smoothness of operation throughout entire range. Check handwheel blacklash.	Pars. 12*b* and 66.
Oilcan points	Wipe clean and oil	Pars. 42 and 43.
Recoil mechanism	Check for excessive oil leakage. Keep respirator in close position.	Pars. 12*e*, 60, and 92.
Recoil slides	Check for rust or damage. Wipe dry and relubricate exposed surfaces.	Pars. 42 and 92.
Sighting and fire control instruments.	Wipe clean and oil exposed bearing surfaces.	Par. 134.
Tires and wheels	Check condition and air pressure.	Par. 103.
Traversing mechanism ..	Check for smoothness of operation throughout entire range. Check handwheel backlash.	Pars. 12*c* and 67.
Monthly Service		
All grease fittings	Wipe clean and lubricate	Pars. 42 and 43.
Range quadrant and telescope mounts.	Clean and oil	Par. 134.
Before Traveling		
Howitzer matériel in general.	Prepare for traveling. Check for loose bolts or parts.	Pars. 23 or 24 and 50.
Blackout lighting system.	Check installation and operation..	Par. 23.
Covers	Check installation	Par. 23 or 24.
Drawbar and pintle	Check engagement and safety pin.	Par. 23.
Hand brakes	Check adjustment	Par. 109.
On-carriage equipment and tool chests.	Check installation and loading...	Par. 23 or 24.
Traveling locks and latches.	Check for proper alinement and engagement.	Par. 23 or 24.
Wheels and tires........	Inspect general condition and air pressure.	Par. 103.

54. Battery (Second Echelon) Maintenance Schedules

a. In general, the battery mechanic is issued necessary tools and either performs or supervises all authorized disassembly, maintenance, or adjustments pertaining to the wheel bearings, hubs, brake drums, and the equilibrator, and supervises the removal of the barrel group and the recoil mechanism for cleaning and painting.

b. Battery service includes a systematic check to see that all crew maintenance (par. 53) has been properly performed at the prescribed intervals and that the matériel is in the best operating condition possible.

Point	Preventive maintenance	Detailed instructions
	Monthly Service	
Howitzer	Remove barrel group. Check bearing strips for wear or scoring.	Par. 71.
Recoil mechanism and respirator.	Remove and clean. Paint where necessary.	Pars. 52, 89, and 92.
Recoil slides	Clean entire length of slides, inspect for damage, and lubricate.	Pars. 42, 47, and 92.
Equilibrator	Check adjustment and general condition.	Pars. 96 and 97.
	Six-Month Service	
Wheel bearings	Clean, repack, and adjust	Pars. 42, 101, and 103.

Section XIII.   MALFUNCTIONS AND CORRECTIONS

55. General

A malfunction is an improper or faulty action of some component part of the weapon, or defective ammunition that may result in failure to fire or other stoppage or damage to the weapon. The following paragraphs tell how to determine the cause of the malfunction and how to remedy it.

56. Failure to Fire

a. If the howitzer fails to fire, first check to see that it is completely in battery, that the breechblock is completely closed, and that the lanyard and firing shaft are functioning; otherwise, the firing mechanism may not rotate the trigger shaft sufficiently to actuate the firing lock, or the firing pin may strike the cartridge case instead of the primer.

b. Attempt to fire the piece twice more. If it still does not fire, wait 60 seconds and then open the breech to extract the cartridge case. Should the complete round be extracted, separate the projectile from the cartridge case.

c. Immediately glance at the primer in the base of the cartridge case and if the indent is normal, throw the cartridge case clear of all personnel due to the continued possibility of a hangfire. Reload with another cartridge case having the same zone charge and resume firing.

d. If the indent on the primer is light or if there is no indent, indication is given that the firing lock is not functioning properly. Remove the firing lock (par. 75*b*), replace it with the spare firing lock, reload, and resume firing.

e. If the malfunction was caused by a defective firing lock, correct it as follows:

(1) If there is no indent on the primer, either the firing pin is broken or deformed or the firing spring or sear spring is weak or broken. Disassemble the malfunctioning firing lock (par. 83) and replace any defective parts.

(2) If the indent on the primer was light or if it fired only after several percussions, indication is given that the firing lock parts are not working freely or that the firing spring is weak. Disassemble the firing lock, clean all parts, and remove any corrosion or burs or foreign matter. Replace firing spring if weak or damaged.

(3) Apply a light film of oil (par. 42) and reassemble (par. 84).

57. Failure to Extract

a. If when the breech is opened, the fired cartridge case is not automatically extracted, remove the cartridge case as outlined in paragraph 20.

b. Remove and examine the extractor (par. 75). Replace, if broken or worn excessively.

c. If extractor is not defective, then examine the removed cartridge case for undersize rim not engaged by the extractor. Or, if cartridge case was removed with difficulty, examine it for bulges or damages and examine the chamber for fouling or foreign matter. Clean chamber, if necessary (par. 47*b*).

58. Seized Breechblock

If breechblock is seized and can neither be closed nor opened without using undue force, inspect for removable obstructions or other cause and correct, if possible. If malfunction cannot be corrected, report seizure to ordnance maintenance personnel.

59. Oil Index Fails to Move

a. If when establishing the oil reserve (par. 88), the oil is pumped against evident pressure and the oil index fails to move from the bottom of the recess, indication is given that the oil index is broken, stuck, or the packing is too tight.

b. Drain off all the reserve oil (par. 88). Screw in 1½ fills of the oil screw filler and tap the index lightly during this operation. If this does not free the index, notify ordnance maintenance personnel.

60. Excessive Leakage of Recoil Oil

If more than a normal leakage of a few drops of dark oil appears around the recoil stuffing box head, oil index, or the filling plug hole, notify ordnance maintenance personnel.

61. Howitzer Returns to Battery with Shock

a. If the howitzer "slams" into battery, it may be due to improper adjustment of the respirator. Increase the buffing action of the respirator (par. 17*e*).

b. There may be an excess oil reserve caused by incorrect filling or by expansion of the recoil oil due to heat from rapid firing. Establish the correct oil reserve (par. 88). If the excess is due to overheating, allow the howitzer to cool before continuing to fire or drain off a slight amount of recoil oil.

c. If the howitzer still returns with excessive shock, notify ordnance maintenance personnel.

62. Howitzer Fails to Return to Battery

a. The respirator may be improperly adjusted. Decrease the buffing action of the respirator (par. 17*e*).

b. There may be deficient oil reserve. Reestablish correct oil reserve (par. 88).

c. Sleigh rails or cradle guides may be dirty or lack lubrication. Remove sleigh (par. 89), clean thoroughly, and lubricate recoil slides. Check for damaged rails, guides, or piston rod. Notify ordnance maintenance personnel of damaged parts.

d. The gas pressure may be low. This is indicated if the oil index shows deficiency after being properly filled, if the oil has an emulsified appearance when released, if the oil does not spurt out, or if the oil screw filler works easily when reestablishing the oil reserve. Report the malfunction to ordnance maintenance personnel.

63. Howitzer Counterrecoil Action Is Jerky

a. If the counterrecoil action is not smooth, indication is given that the sleigh sliding surfaces may be dirty, corroded, lack lubrication, or that air or water has been introduced into the recoil oil.

b. Remove sleigh (par. 89), remove any corrosion or obstructions, clean, and lubricate all sliding surfaces (par. 42).

c. If oil is foamy or has air bubbles or water in it (par. 93) when oil reserve is released, the entire oil supply must be drained and refilled by ordnance maintenance personnel.

64. Howitzer Slides Out of Battery When Elevated

This condition will occur if there is insufficient oil reserve or nitrogen pressure. Drain off oil reserve, and reestablish correct reserve (par. 88). If trouble is not remedied, notify ordnance maintenance personnel.

65. Howitzer Overrecoils or Underrecoils

a. If the howitzer overrecoils, or consistently recoils its maximum distance (par. 21*c*), drain off the oil reserve and reestablish the correct reserve (par. 88). If trouble persists, notify ordnance maintenance personnel.

b. If the howitzer underrecoils (par. 21*c*), check for correct oil reserve. However, if the recoil oil is very cold, the recoil action will normally be sluggish. The recoil will become normal after firing warms up the oil. Check the sleigh and recoil slides for cleanliness and proper lubrication.

66. Howitzer Is Difficult to Elevate or Depress

a. If the elevating handwheels require excessive force to operate them, check for and remove any obstructions from the elevating arcs, pinions, and exterior parts of the elevating mechanism and equilibrator. Clean and lubricate the elevating mechanism. If necessary, adjust the equilibrator as outlined in paragraph 96. If trouble persists, notify ordnance maintenance personnel. Disassembly and lubrication of the equilibrator bearings, cradle trunnions, and elevating gear mechanisms is an ordnance responsibility.

b. If handwheel backlash or free play is in excess of ⅛ turn, notify ordnance maintenance personnel.

67. Howitzer Is Difficult to Traverse

a. If the traversing handwheel requires excessive force to operate it, clean and oil the worm and rack, and lubricate the traversing mechanism and pintle pin bushings (par. 42). Check for and remove any obstructions. If trouble persists, notify ordnance maintenance personnel.

b. If handwheel backlash or free play is in excess of ⅛ turn, notify ordnance maintenance personnel.

68. Hand Brakes Do Not Hold Wheels

If braking action is weak or if brakes drag, adjust brakes properly as outlined in paragraph 109. Remove the hub (par. 99) and check for and remove mud or rust from the brake drum if present. For replacement of excessively worn brake lining or if hand brake cam shaft is seized due to lack of lubrication of brake cam shaft bearing, notify ordnance maintenance personnel.

69. Wheels Loose on Hubs or Spindles

Check wheel stud nuts for tightness. Check proper mounting of hubs for right- and left-hand threaded studs (par. 98). Check adjustment of wheel bearings (par. 101). If bearings are worn or damaged so that proper adjustment cannot be made, notify ordnance maintenance personnel.

Section XIV. BARREL GROUP

70. General

a. The barrel group consists of a barrel assembly (tube assembly), breech ring, and howitzer locking ring. It is mounted on the recoil sleigh assembly.

b. The purpose of the barrel or tube is to house the complete round of ammunition and to direct the projectile when fired. It is rifled to rotate the projectile to aid in maintaining direction and to prevent tumbling in flight.

c. The purpose of the breech ring is to house the breech mechanism (par. 74). The top surface of the breech ring is provided with two leveling plates upon which the gunner's quadrant is placed when checking adjustment of sighting and fire control instruments (par. 128).

71. Removal of Barrel Group

a. GENERAL. The barrel group is removed from the recoil sleigh assembly for painting the under surfaces and to facilitate the removal of the recoil mechanism (par. 89), for cleaning and maintenance of the recoil slides and sleigh rails, or for painting of the recoil cylinder. Methods No. 1 and 2 (*b* and *c* below) are applicable to the field carriage. Method No. 3 (*d* below) is applicable to the motor carriage.

Caution. Due to the weight of the barrel assembly and breech ring adequate precautions to prevent injury to personnel or damage to the matériel must be taken.

b. METHOD No. 1. (1) Check to see that the hand brakes are set.

(2) Remove the breech mechanism as outlined in paragraph 75.

(3) Depress the howitzer sufficiently to compress the equilibrator springs enough to insert a 3" x 3" x 12" wooden block between the equilibrator heads as shown in figure 52. Make sure that the ends of the wooden block are cut squarely. Elevate the howitzer until the block takes up the weight of the rear portion of the cradle. The insertion of this block prevents undue stress or damage to the elevating mechanism parts when the heavy unbalanced weight of the breech ring is moved back to the rear end of the cradle.

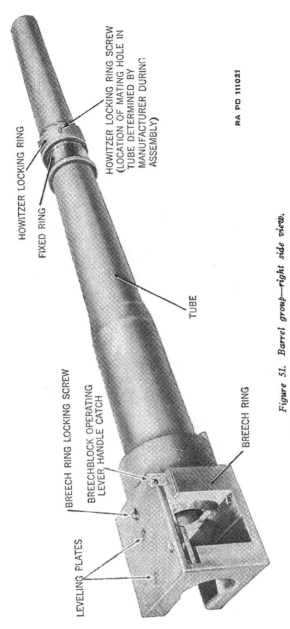

HOWITZER LOCKING RING

FIXED RING

HOWITZER LOCKING RING SCREW
(LOCATION OF MATING HOLE IN
TUBE DETERMINED BY
MANUFACTURER DURING
ASSEMBLY)

RA PD 111021

TUBE

BREECH RING LOCKING SCREW

BREECHBLOCK OPERATING
LEVER HANDLE CATCH

BREECH RING

LEVELING PLATES

Figure 51. Barrel group—right side view.

Figure 52. Blocking equilibrator for removal of barrel assembly and breech ring.

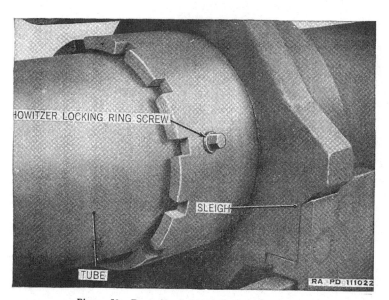

Figure 53. Removing howitzer locking ring screw.

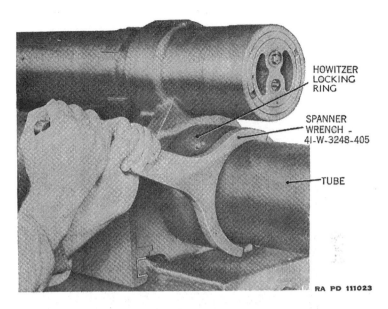

HOWITZER
LOCKING
RING

SPANNER
WRENCH -
41-W-3248-405

TUBE

RA PD 111023

Figure 54. Removing howitzer locking ring.

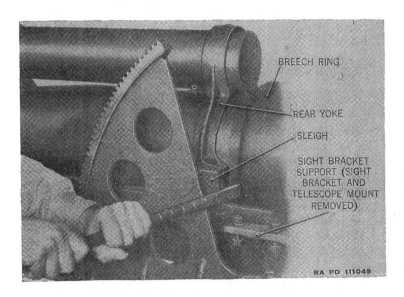

BREECH RING

REAR YOKE

SLEIGH

SIGHT BRACKET
SUPPORT (SIGHT
BRACKET AND
TELESCOPE MOUNT
REMOVED)

RA PD 111049

Figure 55. Prying barrel assembly and breech ring from sleigh with pinch bar.

81

(4) Loosen the howitzer locking screw (fig. 53) and unscrew and remove the howitzer locking ring (fig. 54) using wrench (41–W–3248–405).

(5) The tube is a tight fit in the rear sleigh yoke. Insert a pinch bar between the breech ring and the rear yoke (fig. 55) and pry the barrel assembly and breech ring rearward; or bunt the muzzle end of the tube with a heavy timber until the barrel assembly and breech ring has moved rearward approximately 6 inches and the tightly fitting rear sleigh yoke bearing surface is cleared.

(6) Insert a suitable timber into the muzzle end of the bore and guide the barrel so that the fixed ring does not catch on the sleigh yokes. Insert a suitable timber through the breech ring to support the breech end. Slide and carry the barrel assembly and breech ring to the rear until another carrying timber can be placed under the tube behind the fixed ring. Carefully carry the barrel assembly and breech ring to the rear and place it on wooden blocks to prevent damage to the bronze bearing strips on the bottom of the breech ring.

c. METHOD No. 2. This method requires the use of suitable stable blocking or a chain hoist and a level working area large enough to roll the carriage forward at least 6 to 10 feet.

(1) Place the howitzer at approximately zero elevation.

(2) Following the procedure of *b*, (1), (2), (4), and (5) above.

(3) Slide the barrel assembly and breech ring to the rear until the breech ring extends approximately 4 or 5 inches off the rear end of the cradle. Guide the fixed ring of the tube through the sleigh yokes with a timber inserted into the muzzle end of the bore.

(4) Place a sawhorse or other stable blocking under the breech ring, making sure that it will not sag or sink into the ground when the heavy weight comes upon it, or use a chain hoist, if available.

(5) Release the brakes and roll the carriage forward, at the same time lifting on the timber in the muzzle to guide the tube through the yokes. If necessary, manipulate the elevation of the cradle as required to prevent binding. Even slight maneuvering of the elevating mechanism at this point will not be possible unless the rear of the cradle is lifted to compensate for the unbalanced weight of the barrel assembly and breech ring.

(6) Place a sawhorse or other stable blocking under the fixed ring of the tube and roll the carriage clear.

d. METHOD No. 3. This method requires the use of a chain hoist or other similar hoisting equipment.

(1) In order to provide clearance for removal of the barrel assembly and breech ring when mounted in the motor carriage, it is necessary to depress the howitzer fully and to lower the armor plate flap on the rear wall of the crew compartment.

(2) Follow the procedure of *b*, (2), (4), and (5) above.

(3) Use a chain hoist or similar equipment and very carefully move the barrel assembly and breech ring out of the recoil sleigh assembly. Guide the fixed ring through the sleigh yokes by means of a timber inserted into the muzzle end of the bore. When the fixed ring is clear of the rear yoke, pass another cable or sling around the tube at that point and lift the barrel assembly and breech ring clear of the matériel. Use suitable burlap packing at the suspension points to prevent damage to the parts.

(4) Carefully lower the barrel assembly and breech ring onto blocking to prevent damage to the brass strips on the underside of the breech ring.

72. Installation of Barrel Assembly

a. Sparingly grease the two bearing surfaces of the tube at the points where it is supported by the front and rear sleigh yokes.

b. Install the barrel assembly and breech ring, using suitable timbers or hoists to lift it. Carefully guide the tube through the sleigh yokes to prevent damage to painted surfaces. Bunt the barrel assembly and breech ring into place with a heavy timber.

c. Carefully chcek the mating threads of the howitzer locking ring and the tube and remove any nicks or burs. Lubricate the threads with light graphited grease. Screw the locking ring onto the tube (fig. 54), making sure that the recess drilled into the tube and the hole in the locking ring are in alinement. Screw in the howitzer locking ring screw (fig. 53) and tighten.

d. Install the breech mechanism (par. 76).

e. If a wooden block was used to block the equilibrator (par. 71b(3)), depress the howitzer tube and remove the block.

73. Maintenance of Barrel Group

a. Clean and oil the bore and chamber as prescribed in paragraph 47 at intervals as outlined in the schedule in paragraph 53.

b. Inspect the painted surfaces for any damaged areas and repaint, if necessary, as outlined in paragraph 52. Do not paint over the leveling plates.

c. Inspect the bore for raised, flattened, chipped, or stripped lands, gouges, or other damage. Notify ordnance maintenance personnel for correction of any deficiencies.

d. Decoppering of the bore by using arms is not authorized. A decoppering action will occur, however, whenever reduced charge rounds are fired after a series of normal charge rounds. The gradual coppering action again proceeds whenever firing of normal charge rounds is resumed. If coppering is excessive to the degree where insertion of the round into the chamber becomes difficult, notify ordnance maintenance personnel for correction.

e. The determination of serviceability and the replacement of unserviceable tubes is a function of ordnance maintenance personnel. The accuracy life of the tube, however, depends largely upon the care taken by the battery personnel in its maintenance, the type of rounds fired, the zone charge, and rate and duration of firing. Experience indicates that the tube used in this howitzer, providing it has not been accidentally damaged and has been properly maintained, can fire 20,000 equivalent full charge rounds before erosion progresses to the point where condemnation is warranted.

f. If the retaining screw for the breechblock operating handle catch should accidentally work loose, it will be necessary to stake it in place. Grind two notches about $\frac{1}{16}$ inch deep in the screw seat in the catch, since the catch is too hard to cut with a chisel or file. Insert the catch in its dovetail slot in the breech ring, screw in the retaining screw, and stake it in place.

Section XV. BREECH MECHANISM

74. General

a. The breech mechanism (fig. 56) is housed in the breech ring and is composed principally of the breechblock, the breechblock operating lever, and the extractor. In addition, it houses the firing lock and trigger shaft, which are considered components of the firing mechanism. These are covered in detail in section XVI, except those operations which are essential for the assembly and disassembly of the breech mechanism and covered in this section.

b. The purpose of the breech mechanism is to open the breech, so that a round of ammunition can be inserted, and to close the breech so that the round can be fired.

c. The breechblock is a horizontal-sliding-wedge type and is manually operated by means of the breechblock operating lever. When the lever is unlatched and rotated to the rear, it cams the breechblock horizontally to the right, thereby opening the breech. With the breech open and a round inserted in the chamber, the breechblock is cammed back into the breech ring when the operating lever is manually rotated forward. The front face of the breechblock is beveled and, as the breech is closed, the beveled face contacts the base of the cartridge case and seats it in the chamber.

d. When the breech is completely closed, the breechblock operating lever engages and is locked in the closed position by a catch located in the upper right corner of the breech ring.

e. The extractor is seated in the right side of the breech ring under the breechblock. As the breech is closed, a camming groove in the breechblock rotates the extractor about its seat in the breech ring until the lip

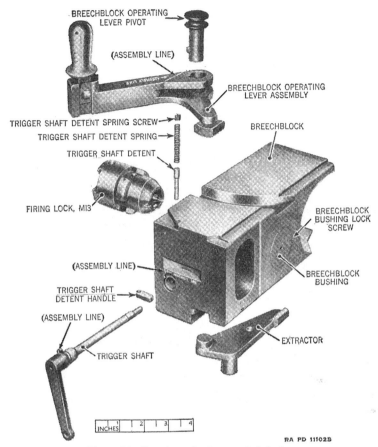

BREECHBLOCK OPERATING LEVER PIVOT

(ASSEMBLY LINE)

BREECHBLOCK OPERATING LEVER ASSEMBLY

TRIGGER SHAFT DETENT SPRING SCREW

TRIGGER SHAFT DETENT SPRING

TRIGGER SHAFT DETENT

BREECHBLOCK

FIRING LOCK, M13

BREECHBLOCK BUSHING LOCK SCREW

(ASSEMBLY LINE)

TRIGGER SHAFT DETENT HANDLE

BREECHBLOCK BUSHING

(ASSEMBLY LINE)

EXTRACTOR

TRIGGER SHAFT

INCHES 1 2 3 4

RA PD 111028

Figure 56. Breech mechanism—exploded view.

end of the extractor is in a recess in front of the rim of the cartridge case. When the breech is opened, the extractor lip, which is in engagement with the rim of the cartridge case, is cammed sharply to the rear, thereby extracting the case from the chamber and ejecting it from the howitzer.

75. Removal of Breech Mechanism

a. Rotate the trigger detent handle out from its recess in the right side of the breechblock and raise it to its uppermost position. While holding it in this position, grasp and withdraw the trigger shaft from the breechblock (fig. 57).

b. Rotate the firing lock 1/6 of a turn in either direction (fig. 58), and withdraw it from the breechblock.

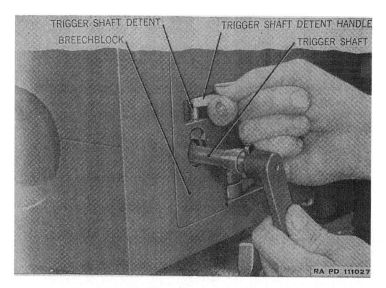

Figure 57. Removing trigger shaft from breechblock.

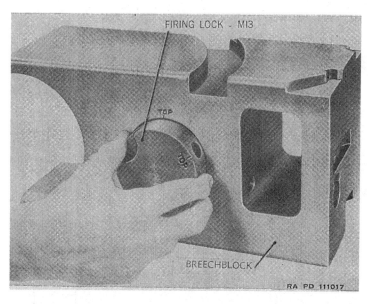

Figure 58. Removing firing lock from breechblock.

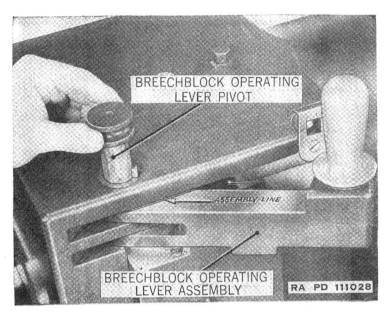

Figure 59. Removing breechblock operating lever pivot.

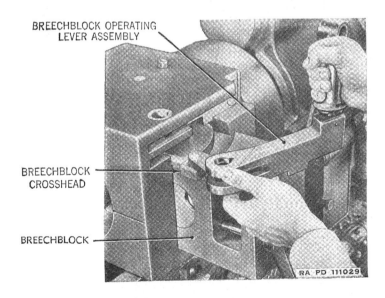

Figure 60. Removing breechblock operating lever.

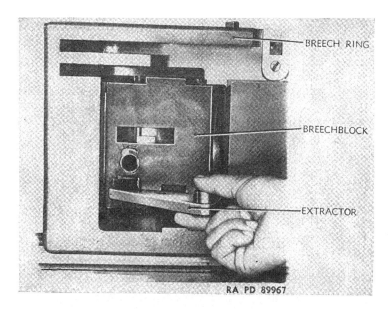

BREECH RING

BREECHBLOCK

EXTRACTOR

RA PD 89967

Figure 61. Removing extractor (breechblock to extreme left).

c. Open the breech to the point where the assembly line on the breech-block operating lever coincides with the side of the breech ring (fig. 59), and lift out the breechblock operating lever pivot.

d. Slide the breechblock to the right and remove the breechblock operating lever when the crosshead has cleared the breech ring (fig. 60).

e. Slide the breechblock to the left only far enough to clear the extractor. Rotate the extractor out of its seat in the rear face of the chamber and lift the extractor from the breech ring (fig. 61).

f. Slide the breechblock to the right and remove it from the breech ring.

76. Installation of Breech Mechanism

a. Insert the breechblock into the breech ring from the right side and slide it to the left only far enough to clear the recess for the extractor lip in the rear face of the chamber. Insert the long trunnion of the extractor into its seat in the breech ring and rotate the extractor forward until the lip of the extractor is in its recess against the rear face of the chamber.

b. Slide the breechblock to the right, making sure that the short trunnion on the extractor engages in its slot in the bottom of the breechblock. When the curved camming groove in the top of the breechblock is clear of the breech ring, insert the crosshead of the breechblock operating lever into the groove as shown in figure 60. Slide the breechblock and breechblock operating lever to the left until the assembly line on the operating lever coincides with the side of the breech ring (fig. 59). Aline the key on the breechblock operating lever pivot with the keyway in its hole in the breech ring (fig. 59) and seat the pivot (fig. 17). Close the breech.

c. Insert the firing lock into its recess in the breechblock so that the alining mark is to the top. Then rotate the firing lock ⅙ of a turn in either direction till the lugs enter their recesses. Push the lock in until its rear surface is flush with the rear face of the breechblock, and rotate it back until the alining marks on the lock and the breechblock coincide.

d. Rotate the trigger detent handle out of its recess in the breechblock and raise it to its uppermost position. While holding it in this position, insert the trigger shaft into its hole in the right side of the breechblock, arm pointing down, and engage its squared end in the firing lock. Secure the trigger shaft in place by releasing the detent handle to allow the detent to engage its groove in the trigger shaft. Rotate the detent handle back into its recess.

77. Disassembly of Breech Mechanism

a. Remove breech mechanism as outlined in paragraph 75.

b. Unscrew and remove the trigger shaft detent spring retaining screw from the top of the breechblock (fig. 62).

c. Tip the breechblock and allow the spring and the detent to drop out of their seat. Take care not to lose the detent handle as it falls from its recess on the right side of the breechblock.

78. Assembly of Breech Mechanism

a. Insert the detent handle into its recess in the right side of the breechblock and aline the hole in the handle with the hole in the breechblock. Drop the trigger shaft detent, with the headed end up, into the hole so that it engages the detent handle (fig. 63).

b. Insert the trigger shaft detent spring into its seat and secure it in place with the retaining screw (fig. 62). Rotate detent handle into its recess in the breechblock.

79. Maintenance of Breech Mechanism

a. Disassemble, clean, and oil the breech mechanism as prescribed in paragraph 47 at intervals as outlined in the schedule in paragraph 53.

b. Repaint any damaged painted areas (par. 52). Do not paint over any operating surfaces.

c. Remove any corrosion, burs, or scored areas with crocus cloth.

d. Replace the extractor if it is broken or damaged so as to fail to extract.

e. Replace the trigger shaft detent spring if it is weak or broken.

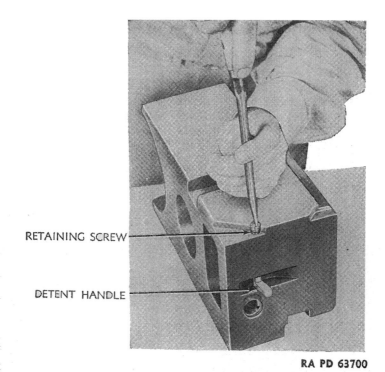

RETAINING SCREW

DETENT HANDLE

RA PD 63700

Figure 62. Removing trigger shaft detent spring retaining screw from breechblock.

f. Keep breechblock operating lever clean and oiled and free of corrosion or burs. If the operating lever handle fails to engage the catch due to a weak or broken spring, notify ordnance maintenance personnel for correction. See paragraph 73*f* if the catch retaining screw accidentally becomes loose.

Section XVI. FIRING MECHANISM

80. General

a. The firing mechanism is composed principally of the firing lock and the trigger shaft, which are housed in the breechblock (fig. 56) and of the trigger shaft pawl, the firing shaft, and the lanyard, which are bracketed on the right side of the cradle (fig. 22).

b. The purpose of the firing mechanism is to fire the round of ammunition.

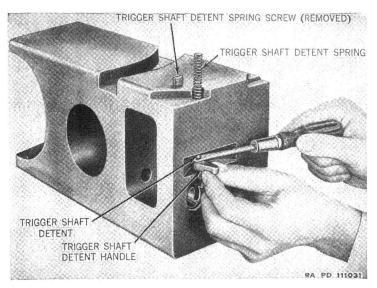

Figure 63. Installing detent assembly into breechblock.

c. As the lanyard is pulled outward, it draws the firing shaft to the rear. Attached to the forward end of the firing shaft is a pawl which contacts and moves the arm of the trigger shaft (fig. 22) to the rear. An inclosed spring returns the firing shaft and pawl to the forward position when the pull on the lanyard is released.

d. The trigger shaft is housed in the breechblock and the squared end of the shaft is engaged with the trigger fork in the firing lock (A, fig. 64). When the arm on the trigger shaft is moved to the rear, the shaft is rotated clockwise and, in turn, its squared end rotates the trigger fork in the firing lock forward (B, fig. 64).

e. The firing lock, M13 is the continuous-pull, safety type, which is not cocked except at the instant before firing.

(1) As the trigger fork is rotated forward, it forces the firing pin holder sleeve forward, thereby compressing the firing spring (B, fig. 64). The sleeve continues to be forced forward until it trips the sear. This releases the firing spring holder and allows the compressed firing spring to expand and snap the firing pin forward (C, fig. 64). The firing pin strikes and detonates the primer in the cartridge case, which in turn ignites the propelling charge.

(2) When the pressure on the trigger fork is released, the firing spring continues to expand with equal force forward and to the rear. The forward pressure is applied to the middle rear surface of the trigger fork by the "T" on the firing pin holder (D, fig. 64). The rearward pressure

91

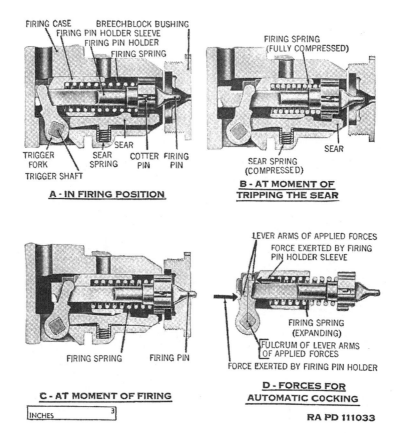

Figure 64. Firing lock M13—sectional view.

is applied to the top front surface of the trigger fork by the holder sleeve (D, fig. 64). Since the rearward pressure exerted to the top front surface acts on a longer lever arm than that of the forward pressure, the trigger fork rotates rearward and moves the holder rearward until the sear again engages the holder as shown in A, figure 64.

81. Removal of Firing Mechanism

a. Grasp the lanyard handle and push the lanyard cord (or strap) through the handle far enough to untie the knot. Withdraw the lanyard handle from the cord. Remove the S-hook from the bracket. Grasp the hook and pull the cord free through the pulleys (fig. 22).

b. Remove the trigger shaft and firing lock, if necessary, as outlined in paragraph 75*a* and *b*.

82. Installation of Firing Mechanism

a. Thread the lanyard cord through the pulley on the rear end of the firing shaft and then through the pulley of the lanyard bracket as shown in figure 22. Slip the lanyard handle over the end of the cord and tie a knot in the cord. Pull the lanyard handle down over the knot. Engage the S-hook in the eye of the bracket.

b. If firing lock and trigger shaft have been removed, install as outlined in paragraph 76.

83. Disassembly of Firing Lock M13

a. Remove the trigger fork by grasping the firing case, with trigger fork down, and pressing the firing pin against a solid surface (fig. 65). The trigger fork will fall free of the firing lock into the hand. An alternate method is to pry the trigger fork out of the firing case with a screw driver, first through the trigger shaft hole (fig. 66), and then from the outside of the case.

b. Press the front end of the sear out of engagement with the firing pin holder; at the same time pry the assembled sleeve and holder forward (fig. 67) until they can be grasped and pulled from the firing case. Shake out the sear and sear spring.

c. To disassemble the firing pin holder and sleeve, grasp the front end of the firing pin holder sleeve and place the lower rear end of the sleeve

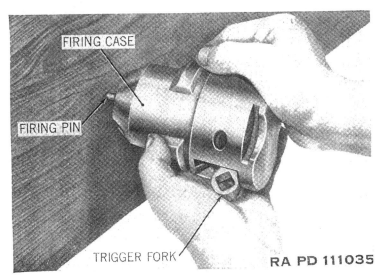

FIRING CASE

FIRING PIN

TRIGGER FORK

RA PD 111035

Figure 65. Removing trigger fork from firing lock.

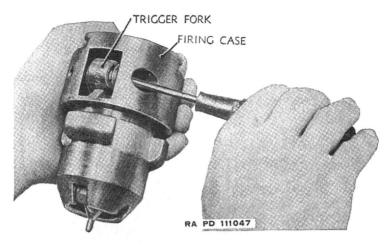

TRIGGER FORK

FIRING CASE

RA PD 111047

Figure 66. Removing trigger fork from firing lock—alternative method.

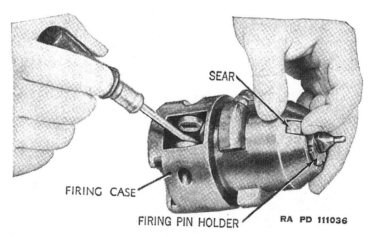

SEAR

FIRING CASE

FIRING PIN HOLDER

RA PD 111036

Figure 67. Removing firing pin holder and sleeve from firing case.

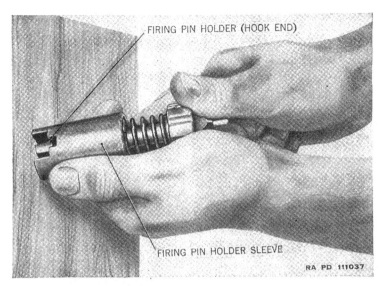

FIRING PIN HOLDER (HOOK END)

FIRING PIN HOLDER SLEEVE

RA PD 111037

Figure 68. Removing sleeve from firing pin holder.

against a solid surface. Push the firing pin holder down to unhook it from the sleeve (fig. 68). Allow the holder to recede out of the sleeve, freeing the spring.

d. Remove the cotter pin from the firing pin holder and unscrew the firing pin.

84. Assembly of Firing Lock M13 (fig. 69)

a. Screw the firing pin into the firing pin holder, insert the cotter pin and spread the ends carefully so that they will not rub against the firing case. Assemble the firing spring over the holder and the sleeve over the spring and holder (fig. 68) and, pushing the rear of the holder against a solid surface, compress the spring enough to hook the T of the holder in the T-slot of the sleeve.

b. Insert the sear spring into its seat in the bottom of the firing case, using a screw driver between two coils of the spring (fig. 70). Assemble the sear into the case, so that the sear spring stud enters the sear spring.

c. Press the sear down with a screw driver inserted through the hole in the firing case (fig. 71) and insert the assembled firing pin holder and sleeve into the case, with the flat portion of the sleeve and the sear notch of the holder downard, so that they will engage the sear. Hold the sear from slipping backward, withdraw the screw driver, and push the holder back until it is latched by the sear.

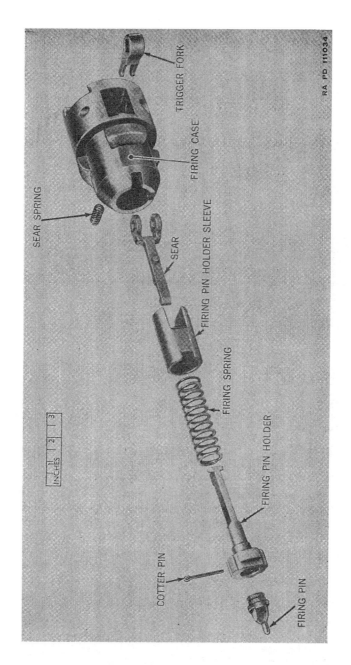

INCHES
1 2 3

SEAR SPRING

TRIGGER FORK

FIRING CASE

SEAR

FIRING PIN HOLDER SLEEVE

FIRING SPRING

FIRING PIN HOLDER

COTTER PIN

FIRING PIN

RA PD 111034

Figure 69. Firing lock M13—exploded view.

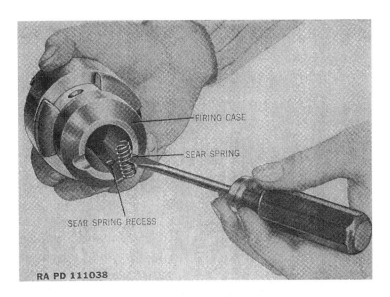

Figure 70. Inserting sear spring into firing case.

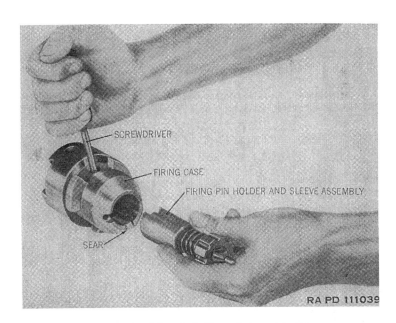

Figure 71. Inserting firing pin holder and sleeve into firing case.

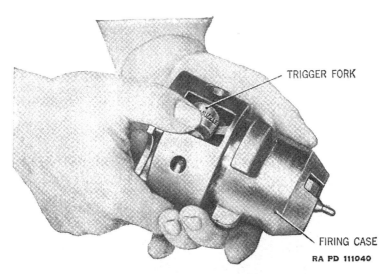

TRIGGER FORK

FIRING CASE

RA PD 111040

Figure 72. Inserting trigger fork into firing lock.

d. Insert the trigger fork into the opening in the bottom of the case, having the part marked "MUZZLE FACE" toward the front (fig. 72). Push the trigger fork until it snaps into position.

85. Maintenance of Firing Mechanism

a. Keep parts clean and properly lubricated (pars. 42 and 47). Remove corrosion or burs and smooth roughened bearing surfaces with crocus cloth.

b. Repaint any damaged painted areas (par. 52).

c. If the firing pin is worn, damaged, or deformed, replace it. If the sear spring or firing spring are weak or broken, replace them. If the firing pin holder or cotter pin are worn or damaged, replace them. Replace complete firing lock for irreparable deficiencies in other parts not issued.

d. If the lanyard cord or strap is weak, frayed, or broken, replace it. Apply neat's-foot oil to keep leather lanyards and straps soft and pliable.

e. If the firing shaft spring becomes weak or broken, notify ordnance maintenance personnel.

Section XVII. RECOIL MECHANISM

86. General

a. The recoil mechanism (figs. 73 and 74) is composed of the recoil sleigh assembly, the recuperator cylinder, and the recoil cylinder. These parts together with the barrel group, which is securely locked to the sleigh assembly, are known as the recoiling parts. The recoil mechanism serves to absorb the energy and shock of firing by gradually checking and stopping the rearward movement of the recoiling parts; it returns them into the battery position during counterrecoil; provides proper buffing action to prevent "slamming;" and it holds them in the battery position by the force of compressed nitrogen in the recuperator cylinder. It is of the hydro-pneumatic, constant recoil type, employing a floating piston to separate the recoil oil from the nitrogen gas. A pneumatic respirator in the recoil cylinder provides the counterrecoil buffing action.

b. The recoil sleigh assembly houses and supports the recuperator cylinder, the recoil cylinder, and the barrel assembly. The sleigh includes three yokes which hold the parts together, and two rails which slide on the stationary cradle to guide and support the recoiling parts. The barrel assembly is supported by the front and rear yokes and is secured to the recoil sleigh assembly by a locking ring.

c. The recuperator cylinder contains compressed nitrogen gas, held between the floating piston and the rear head. The front part of the cylinder contains recoil oil, under pressure, filling the regulator body and the space between the floating piston and the regulator. The front head contains a filling valve mechanism and the oil index mechanism.

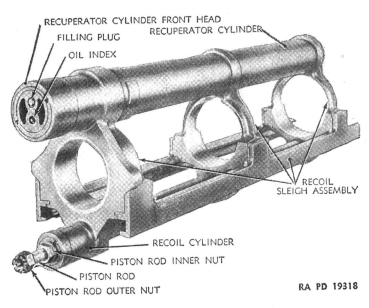

RA PD 19318

Figure 73. Recoil mechanism—front view.

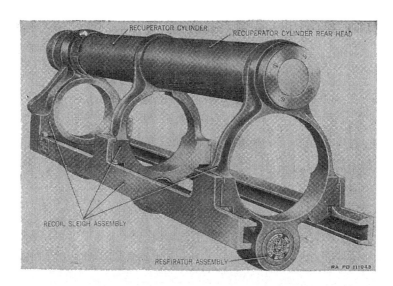

Figure 74. Recoil mechanism—rear view.

d. The recoil cylinder contains recoil oil which can pass back and forth between the recoil cylinder and recuperator cylinder by means of connecting passages. A recoil piston, which is secured to the cradle by a piston rod, holds this oil and separates it from air which fills the rear portion of the cylinder. A stuffing box prevents leakage of oil past the piston rod. The rear head of the cylinder houses the respirator, which permits air to enter the recoil cylinder during recoil and controls the escape of the air through an adjustable valve during counterrecoil.

87. Functioning

a. ACTION IN RECOIL (fig. 75). (1) When the howitzer is fired, the force of the expanding gas propels the projectile out of the bore. This same force also reacts against the breechblock and forces the recoiling parts (par. 86*a*) rearward, except for the recoil piston in the recoil cylinder. The piston is held from recoiling by the piston rod, which is attached to the front end of the cradle by means of the inner and outer nuts. As the sleigh moves back in recoil, the recoil oil in the recoil cylinder is forced through the communication passages in the front yoke of the sleigh into the regulator body of the recuperator cylinder.

(2) The regulator body contains one-way valves and a throttling opening (inset, fig. 75) through which the oil passes, where it acts upon the floating piston diaphragm in the recuperator cylinder, forcing it to the rear, and further compressing the nitrogen behind the floating piston.

(3) As the floating piston moves to the rear, a tapered control rod which is fastened to the diaphragm is drawn through the throttling orifice.

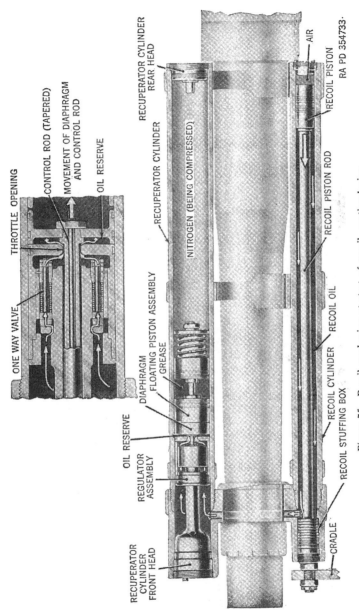

THROTTLE OPENING

CONTROL ROD (TAPERED)

MOVEMENT OF DIAPHRAGM AND CONTROL ROD

OIL RESERVE

ONE WAY VALVE

RECUPERATOR CYLINDER REAR HEAD

RECUPERATOR CYLINDER

AIR

RECOIL PISTON

RA PD 354733.

NITROGEN (BEING COMPRESSED)

RECOIL PISTON ROD

DIAPHRAGM

FLOATING PISTON ASSEMBLY

GREASE

RECOIL OIL

RECOIL CYLINDER

RECOIL STUFFING BOX

OIL RESERVE

REGULATOR ASSEMBLY

CRADLE

RECUPERATOR CYLINDER FRONT HEAD

Figure 75. Recoil mechanism at start of recoil—sectional view.

101

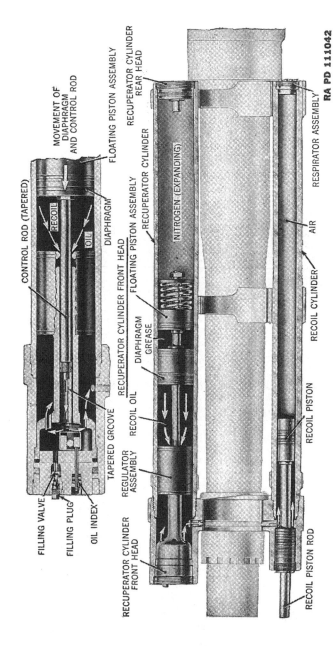

Figure 76. Recoil mechanism during counterrecoil—sectional view.

FILLING VALVE

FILLING PLUG

OIL INDEX

RECUPERATOR CYLINDER
FRONT HEAD

REGULATOR
ASSEMBLY

TAPERED GROOVE

RECOIL OIL

DIAPHRAGM
GREASE

RECUPERATOR CYLINDER FRONT HEAD

FLOATING PISTON ASSEMBLY

RECUPERATOR CYLINDER

NITROGEN (EXPANDING)

CONTROL ROD (TAPERED)

RECOIL

OIL

DIAPHRAGM

MOVEMENT OF
DIAPHRAGM
AND CONTROL ROD

FLOATING PISTON ASSEMBLY

RECUPERATOR CYLINDER
REAR HEAD

RESPIRATOR ASSEMBLY

AIR

RECOIL CYLINDER

RECOIL PISTON

RECOIL PISTON ROD

RA PD 111042

The area through which oil can flow is thus reduced gradually to the point where the remaining energy of recoil is unable to force the oil to the rear. At that time, since the resistance is equal to the force, further motion is not possible and the recoiling parts are brought to rest.

(4) The energy of the recoiling parts is principally exhausted in the work forcing the oil through the orifice in the regulator. Some of the energy, however, is spent in compressing the nitrogen gas, which has a normal pressure of approximately 1100 pounds per square inch, and in overcoming the combined friction of all the moving parts.

b. ACTION IN COUNTERRECOIL. (1) When the recoiling parts are brought to rest at the end of recoil, the unbalanced force of the greatly compressed nitrogen gas forces the floating piston and diaphragm forward, pushing the recoil oil back through the regulator into the recoil cylinder against the back of the recoil piston (fig. 76), thereby returning the recoiling parts to battery position.

(2) However, the oil does not return through the channels in the regulator by the way in which it entered. The one-way regulator valves are closed under pressure of the regulator valve springs. The returning oil is diverted to the central bore of the regulator, where another throttling action takes place as the oil passes by the control rod piston through grooves of decreasing depth (inset, fig. 76) cut in the walls of the regulator bore.

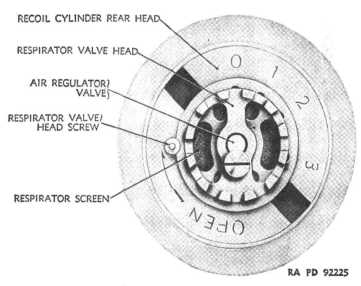

RA PD 92225

Figure 77. Respirator.

(3) By the return of oil past these tapered clearances, the speed of counterrecoil is controlled and reduced so as to return the howitzer to battery slowly.

c. ACTION OF THE RESPIRATOR. (1) The velocity of the counterrecoil is dependent to some extent upon the elevation of the howitzer and the viscosity of the recoil oil. Since the throttling grooves are not adjustable, a pneumatic buffer, the respirator, has been provided to compensate for any changes in the velocity of counterrecoil. This adjustable buffer, which controls the rate at which air can escape from the recoil cylinder during counterrecoil is located in the rear head of the recoil cylinder.

(2) The respirator controls the escape of air by means of an adjustable orifice, which can be set at any one of four positions. This orifice is controlled by a valve which is attached to the head of the respirator. By admitting air freely during recoil and restricting the escape of air during counterrecoil, the respirator provides the additional buffing action which is sometimes necessary to prevent the howitzer from slamming into battery.

d. OIL INDEX AND OIL RESERVE. (1) The recoil mechanism is designed to operate properly when correct recoil oil reserve is forced into the system so as to separate the floating piston diaphragm from the regulator, thereby transmitting the pressure of the nitrogen gas through the floating piston, to the oil column. The pressure of the recoil oil acting on the recoil piston holds the howitzer in the battery position. Lack of the oil reserve may cause the howitzer to fall out of battery at high elevation. Correct recoil oil pressure exists when the end of the oil index indicator rod is flush with the front face of the recuperator cylinder front head, indicating a sufficient amount of reserve oil.

(2) Whenever the amount of reserve oil is less than that prescribed, a rod attached to the diaphragm moves forward with the floating piston diaphragm and actuates the pinion and rack oil index mechanism so as to cause the index indicator rod to recede into the oil index recess, indicating insufficient reserve oil.

(3) However, if the oil reserve pressure is excessive, the oil index is mechanically unable to indicate this excess condition due to the construction of the mechanism indicator. This condition necessitates extreme care in establishing correct oil reserve.

88. Draining and Reestablishing the Oil Reserve

a. To drain the oil reserve, remove the filling plug (fig. 25) with a suitable wrench; then insert the liquid releasing tool into the filling hole and hand tighten it. Using a suitable wrench, further tighten the tool (fig. 78) until the oil resereve spurts out in a stream. If checking the

104

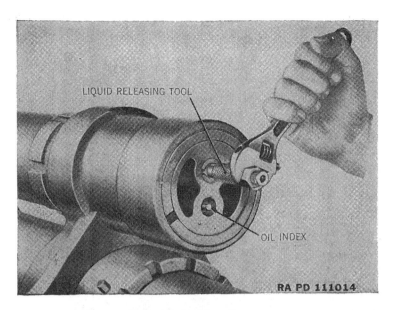

Figure 78. Draining oil reserve.

reserve before firing, drain only enough oil to cause the oil index to recede slightly into the recess, and catch the oil in a suitable receptacle or waste material. For complete draining of the oil reserve, allow the oil to spurt out until the flow stops. Unscrew the liquid releasing tool.

b. Inspect the recoil oil as described in paragraph 93. Use oil prescribed by the War Department Lubrication Order 9–325.

c. To fill the screw-type, hand operated, oil screw (filler) gun, turn the handle counterclockwise until screwed completely back, loosen the locking screw on the head, and remove the handle and head as a unit. Pour oil directly into the barrel of the (filler) gun, avoiding the formation of air bubbles. Replace handle and head as a unit and tighten the locking screw. Remove the cap from the nozzle head, hold the nozzle end up for a minute or two until all the air in the oil has risen to the surface, and purge the gun by turning the handle until no more air bubbles appear on the nozzle end.

d. To reestablish the oil reserve, screw the nozzle of the gun into the filling hole, taking care not to cross the threads. Before tightening, turn the handle and force out any air in the filling hole. Operate the gun with both hands and avoid lateral pressure on, and possible breakage of, the threaded nozzle (fig. 79). When the oil index shows full reserve (approximately 1½ fills), unscrew the gun and install the filling plug.

e. If oil index fails to move against evident pressure, see paragraph

105

RECOIL OIL (FILLER) GUN

OIL INDEX

RA PD 111015

Figure 79. Establishing oil reserve.

59 for corrective action. If it still fails to move, the howitzer may be fired in an emergency until the piece either fails to return to battery or returns with shock, in which case see paragraph 61 or 62.

89. Removal of Recoil Mechanism

a. Remove the barrel group by one of the methods given in paragraph 71.

b. Remove the cotter pin and piston rod outer nut (fig. 73). For the motor carriages, it will be necessary to first remove the armor plate protecting the front end of the cradle. Also remove the recuperator cylinder shield to reduce weight.

c. Slide the sleigh assembly to the rear until it is clear of the cradle, being careful to lift the rear of the sleigh sufficiently to prevent cramping of the sleigh rails and the forward part of the recoil slides, and to prevent the sleigh from dropping onto the rear portion of the recoil slides when it clears the forward part.

Note. The recoil sleigh assembly weights approximately 463 pounds (without armor plate).

d. Place the sleigh assembly on suitable blocking to prevent damage to the recoil cylinder.

90. Installation of Recoil Mechanism

a. Slide the sleigh assembly onto the cradle, taking care to prevent binding or damage to the brass slides of the cradle.

b. Replace the piston rod outer nut and cotter pin. Draw the nut up just tight enough to prevent end play and back it off one castellation. This will allow the piston rod to find its natural position without binding and causing a leak at the stuffing box. For the motor carriages, replace the armor plate on the front end of the cradle and the recuperator cylinder shield, if removed.

c. Install the barrel group (par. 72).

91. Retraction of Howitzer and Sleigh

a. When it is impractical to completely remove the barrel group and the sleigh assembly from the matériel for cleaning the recoil slides and sleigh rails, it is necessary to retract the howitzer as far as possible, using appropriate precautions. Level the tube, block the equilibrator (par. 71*b* (3)) on motor carriages, remove the armor plate from the front end of the cradle, and then remove the cotter pin and piston rod outer nut. Slide the complete barrel and sleigh assemblies toward the rear to the end of the recoil slides or onto stable blocking as far to the rear as clearance will permit.

Note. The combined barrel and sleigh assemblies weigh approximately 1,527 pounds, without the cradle or recuperator cylinder shields.

b. After proper cleaning and maintenance (par. 92), carefully slide the assemblies back into battery position. Replace the piston rod outer nut and cotter pin as prescribed in paragraph 90*b*. Remove the blocking from the equilibrator and reinstall any armor plate which has been removed.

92. Maintenance of Recoil Mechanism

a. Battery maintenance of the recoil mechanism is limited to exterior cleaning (par. 47), painting (par. 52), draining and reestablishing the oil reserve (par. 88), and to checking for improper length of recoil or faulty recoil or counterrecoil (pars. 61 to 65).

b. Keep the recoil slides and sleigh rails clean, well lubricated, and free from burs, scoring, corrosion, or other damage.

c. Every precaution must be taken in servicing the recoil mechanism to keep recoil oil, liquid releasing tool, oil screw (filler) gun, and areas around the filling plug clean and to prevent dust, sand, or dirt from getting inside the finely machined surfaces, as foreign matter may cause irreparable damage.

d. When the weapon is not being fired, keep the valve of the respirator set at "0" (par. 12*e*) in order to keep accumulations of moisture or dust out of the recoil cylinder. Keep respirator as clean as possible. Tools for disassembly of the respirator are not issued to using arms.

93. Care of Recoil Oil

a. Water or foreign matter must not be introduced into recoil oil or the recoil mechanism. Exposure of recoil oil in an open container or partly filled container may result in an accumulation of water, either directly or by condensation of moisture on the sides of the container. Drained recoil oil should not be reused except in emergency.

b. If recoil oil has been exposed to moisture or if it is to be reused in an emergency, it should be tested for water as outlined below.

Note. If in doubt about the presence of water after conducting test, change the oil. Tests for presence of water are not conclusive under all conditions.

(1) Use a clean glass bottle of 1-pint capacity, filled with the recoil oil. Allow the bottle to stand undisturbed for several hours. If water is present, it will sink to the bottom. When the bottle is lightly tilted, drops or bubbles will form. Invert the bottle and hold to the light. Water, if present, may be seen in the form of droplets slowly sinking in the oil. If the oil has a cloudy appearance, it may be ascribed to finely divided particles of water scattered throughout the oil.

(2) Another test for water is to heat a shallow pan of oil to 212° F. (boiling point of water). Water in the oil will appear on the surface as tiny bubbles and will be disclosed by this test when not determinable by the settling test.

(3) Should either of these tests show water, the oil on hand should not be used but should be returned through supply channels for reclamation.

c. Drained recoil oil, if it is to be reused in an emergency, must be strained through clean lintless cloth or linen to exclude foreign matter.

d. Take especial care to preserve the identity and grade of recoil oil. Do not mix recoil oil of different grades or other oil. Do not use recoil oil as a lubricant. Keep it protected from excessive heat.

Section XVIII. CRADLE, CRADLE LOCKS, AND EQUILIBRATOR

94. Description and Functioning

a. The cradle is a trough shaped mechanism (fig. 7) that supports the heavy recoiling parts. It is pivoted on trunnions and with the elevating mechanism, which is attached to it, provides means for elevating and depressing the howitzer. The sighting and fire control instruments are also attached to it. The recoiling parts are connected to the front end of the cradle by inner and outer piston rod nuts.

b. During traveling, the forward weight of the cradle and howitzer is supported by the cradle lock strut (fig. 80), which is hinged to the underside of the cradle and engages the lower strut latch on the carriage frame, thereby preventing chatter on the elevating arcs. The rear end of

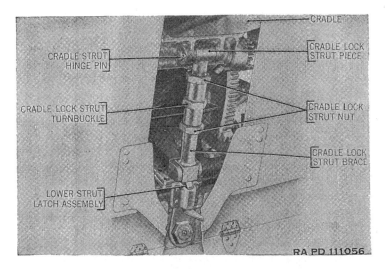

Figure 80. Cradle lock strut in traveling position.

the cradle is supported by the ball-shaped traveling lock shaft pieces which engage sockets in the traveling lock brackets bolted to the trails (fig. 36).

c. The spring type equilibrator connects the rear end of the cradle to the top carriage (fig. 19). The preponderance of the weight supported by the cradle is to the front of the trunnion bearings when the recoiling parts are in battery. The function of the equilibrator is to compensate for the unbalanced weight and reduce the manual effort required to elevate and depress the howitzer.

95. Adjustment of Cradle Lock Strut and Traveling Lock Brackets

a. The traveling lock brackets must be kept in proper alinement and the bolts must be tight in order that the trails can be closed properly. Do not force the trails. A small amount of adjustment is possible by loosening the cap screws securing the bracket to the trails (fig. 36), tightening or loosening the adjusting nut until proper alinement of the socket and the ball-shaped end pieces are secured, and retightening the cap screws.

b. The cradle lock strut can be adjusted, if it is too long or too short to engage the lower strut latch when the trails are closed, by loosening the jam nuts on the strut and turning the turnbuckle (figs. 2 and 80) until the exact proper length is secured. Retighten the jam nuts.

96. Adjustment of Equilibrator

a. Operate the elevating handwheel through the entire range and check for smoothness of operation. If difficult or jerky, inspect the elevating mechanism for obstructions, gummy oil, corrosion, burs, and other possible sources of trouble. Adjust the equilibrator only after it has been positively found to be out of adjustment, and then adjust with care. Do not try to force the elevating handwheels, as damage to the elevating worm and worm gear will occur and necessitate needless ordnance maintenance repair.

b. To adjust the equilibrator, loosen the jam nuts (if provided) on the three equilibrator guide rods, and adjust the tension of the equilibrator spring by tightening or loosening the guide rod nuts with a suitable wrench.

Caution. Take extreme care to adjust the three guide rod nuts evenly. They should be loosened if the howitzer is hard to depress and easy to elevate; and tightened if easy to depress and hard to elevate. Tighten the jam nuts after adjustment is obtained.

c. Some equilibrators use staking instead of jam nuts and the guide rods do not provide sufficient thread space for adjusting. For such units, no adjustment can be made. Notify ordnance maintenance personnel for replacement of equilibrator.

97. Maintenance of Cradle, Cradle Locks, and Equilibrator

a. The brass recoil slide strips on the cradle will be kept clean (par. 47), lubricated (War Department Lubrication Order), and free from scoring or burs. Be sure that piston rod outer nut is properly tightened (par. 90*b*) and that the cotter pin is inserted and spread.

b. Keep the cradle lock strut and traveling lock brackets in proper alinement and adjustment (par. 95). Keep them free from burs or corrosion and repair any damage.

c. Removal of the equilibrator, for cleaning and lubrication of the spring rod needle bearings and the fulcrum bearings, is performed periodically by ordnance maintenance personnel. However, using arms personnel are responsible for general outside cleaning and maintenance of the parts and oiling of the guide rods.

d. The cradle trunnion bearings are disassembled and lubricated periodically by ordnance maintenance personnel only.

Section XIX. WHEEL ASSEMBLY

98. Description and Functioning

The towed field carriages are provided with divided disk and rim type wheels with 9.00 x 20 combat tires and tubes. Tires may have commercial type treads or nondirectional mud and snow treads. A directional tread will not normally be used on a towed vehicle. The tire is mounted on the disk and rim and secured in place by means of the rim ring. A

Figure 81. Removing wheel hub stud nuts.

hinged-type beadlock prevents the tire from slipping around the rim. The wheel is fastened to the hub by stud bolts. Studs for the left side hub have left-handed threads while those on the right side have right-handed threads; however, some carriages were manufactured with right-hand studs only. The hubs roll on tapered roller bearings.

Note. Some towed field carriages are provided with the 7.50 x 24 standard 8-ply tires and wheels instead of the combat tires and wheels. Refer to paragraph 102c below.

99. Removal of Wheel and Hub Assemblies

a. REMOVAL OF WHEEL ASSEMBLY. (1) Set the brakes and loosen the six wheel stud nuts, using the wheel stud nut wrench and handle, while the wheel is on the ground (fig. 81). Jack up the carriage and place blocking under the axle.

(2) Remove the stud nuts and lift the wheel assembly from the hub. It weighs approximately 287 pounds.

b. REMOVAL OF HUB ASSEMBLY. (1) Remove the wheel (*a* above).

(2) Remove the three screws retaining the hub cap, using a screw driver, and remove the hub cap and gasket.

(3) Remove the cotter pin and the wheel spindle castle nut (inset, fig. 82).

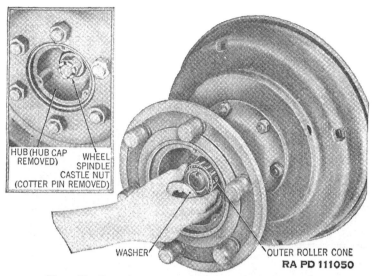

Figure 82. Removing outer cone and washer from wheel spindle.

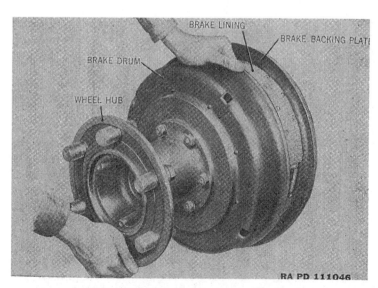

Figure 83. Removing hub and brake drum.

(4) Remove the washer and outer roller bearing (fig. 82), after starting by pulling out on the hub slightly.

(5) Remove the hub and brake drum (fig. 83). Usually the inner roller bearing and oil seal will remain with the hub.

(6) Drive out the oil seal and inner roller bearing from the hub, using a wooden block or brass drift and a hammer. Shift the position of the drift all the way around the hardened surface of the bearing and tap lightly.

(7) Check condition of brake lining and brake mechanism (sec. XX) before installing hub and brake drum.

100. Installation of Wheel and Hub Assemblies

a. INSTALLATION OF HUB ASSEMBLY. (1) After the bearings have been packed (par. 103), install the inner roller bearing and the oil seal in the hub, making sure that the seal is serviceable and has the leather edge facing inward. Use a wooden block or brass drift and tap lightly all around the seal to prevent damage to the soft metal (fig. 84).

(2) Slide the hub and brake drum over the axle spindle, taking care not to damage the oil seal.

Caution. If the hubs are equipped with right- and left-handed thread stud bolts, be sure that the hub with left-handed bolts is installed on the axle on the left side.

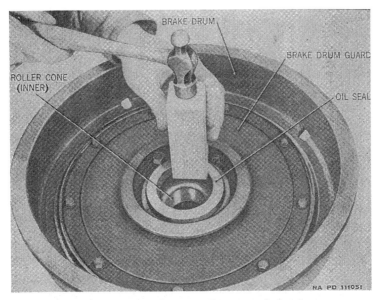

Figure 84. Installing inner roller cone and oil seal.

(3) Slide the outer roller bearing on the spindle into its seat in the hub. Install the washer and wheel spindle castle nut. Do not insert cotter pin until wheel bearings have been adjusted (par. 101).

b. INSTALLATION OF WHEEL ASSEMBLY. Mount the wheel assembly on the hub and secure in place with the six wheel stud nuts, tightening diametrically opposite nuts until all are tight.

101. Adjustment of Wheel Bearings

a. Install hub and wheel assembly (par. 100), leaving the cotter pin out of the wheel spindle castle nut, the hub cap removed, and the wheel jacked clear of the ground.

b. Tighten the wheel spindle castle nut, at the same time rotating the wheel in both directions as the nut is tightened until all the bearing parts are firmly seated. At this point, a drag or resistance to rotation of the wheel will be felt. Back off the nut one-sixth to one-quarter of a turn to the nearest castellation where the cotter pin can be inserted. The wheel should then rotate freely but without any perceptible shake. Test by feeling for slight jar or movement in the hub when trying to shake the wheel to and away from the carriage.

c. Insert and spread the cotter key. Assemble the hub cap and tighten the three hub cap screws.

d. Remove the jack or blocking from under the axle.

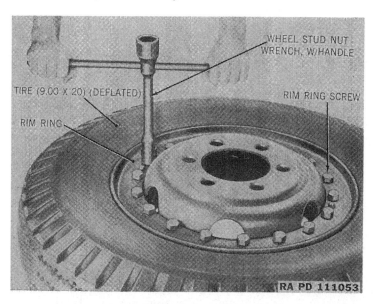

Figure 85. Removing disk and rim ring screws.

102. Changing and Repairing Tires and Tubes

a. DISMOUNTING TIRE AND TUBE. (1) Remove the wheel assembly from the hub (par. 99). *Deflate the tire by removing valve core.* Reinstall valve core.

Caution. It is extremely dangerous to remove a tire without first deflating it.

(2) Remove the 18 disk and rim ring screws (fig. 85), leaving two diametrically opposite screws in place to hold the ring until all the other screws have been removed. Remove the rim ring, using a suitable tire iron if necessary.

(3) Remove the tire and beadlock from the disk and rim by dropping the wheel onto a small wooden block or by knocking the tire loose with a suitable heavy wooden or rubber mallet.

(4) Lay the tire flat on the ground with the valve stem pointing up and unlock the beadlock hinge with a heavy screw driver (fig. 86). Pull the beadlock free, taking care not to damage the valve.

(5) Remove the tube, separating and holding the side walls open with short wooden blocks as necessary.

(6) Repair all the punctures, cuts and bruises in the tube and tire, using cold patches and boots as required. Replace tire or tube if damage cannot be repaired.

b. MOUNTING TIRE AND TUBE. (1) Insert the tube into the tire carefully to avoid twists or wrinkles. If tube is synthetic rubber, which can be identified by a blue or red stripe on inner circumference, special care must be taken to prevent chafing and undue stresses. Dust the tire, tube, and rim with talcum soapstone or lubricate with heavy suds solution of soap flakes. Inflate tube slightly to round it out. Deflate to allow stresses to adjust themselves and then inflate it till the tire side walls are spread sufficiently to receive the beadlock.

(2) Place the beadlock over the valve and insert it in the tire, centering it to fit properly between the beads. Lock the hinge.

(3) Install the tire onto the disk and rim so that the valve stem points out through one of the access holes in the wheel disk.

(4) Install the rim ring, lining up the valve slot, and force the ring down far enough to start a screw near the valve and one directly opposite. Continue to install screws in alternate positions until all 18 screws are in place and tight.

(5) Inflate the tire to the correct pressure and replace valve cap.

c. CHANGING 7.50 x 24 STANDARD TIRE. If the carriage is equipped with 7.50 x 24 standard 8-ply tires and wheels instead of combat wheels (some carriages, which were modified from the M2 model), the procedure for dismounting and mounting is the same as in *a* and *b* above, except that a ridged tire locking ring is used in place of the bolted rim ring and there is no beadlock.

(1) To remove the tire locking ring, first deflate the tire and break the bead loose, using a heavy rubber or wooden mallet. Then insert a heavy screw driver or similar tool into the notch in the tire locking ring and pry the ring out, at the same time pounding downward to release the locking ridge from the rim gutter (fig. 87). Work around the rim, continuing to pry out the locking ring until it can be lifted from the rim.

(2) To install the rim locking ring, start it at one end, forcing the ridge under the gutter and continue to work around the rim until the rim locking ring is locked under the gutter all the way around the rim.

Caution. The rim locking ring is apt to spring out and cause serious injury to personnel if not securely locked under the gutter before air pressure is applied.

Note. Where a 7.50 x 24 standard tire is replaced with the 9.00 x 20, 12-ply combat tire and wheel, it is necessary to also replace the hub.

103. Maintenance of Wheel and Hub Assemblies

a. TIRES. The proper tire pressure under average conditions is 40 pounds per square inch. It may be reduced to 32 pounds per square inch for low speed over soft terrain, when increased flotation is necessary, or increased to 48 pounds per square inch for high speeds over improved hard surface highways. Do not "bleed" tires because the pressure has increased during travel. Keep tires covered when under the direct rays of a hot sun. Probe tires regularly to remove bits of glass, rock, or nails to prevent their working through the rubber and causing damage and slow leaks. Follow the instructions in paragraph 102*b* for care of synthetic rubber inner tubes when changing or repairing the tubes. Keep all petroleum products off rubber as they will deteriorate the rubber rapidly.

b. WHEEL AND HUB ASSEMBLIES. (1) *General.* Keep metal surfaces clean and painted. Stones chip the paint during travel and water and dirt collect under the rim locking ring and in other crevices, causing rapid corrosion. Do not use a metal hammer on the painted surfaces as it will nick the surface and knock off the paint. Keep the wheel stud nuts tight. Although designed to work tighter instead of loose in normal operation, a wheel that has had nuts turned hand-tight only will rapidly wear out the studs and fly off the hub with consequent unnecessary damage to matériel and danger to personnel. Keep wheel bearings in adjustment. Too tight an adjustment will cause the wheel to run hot, score and chip the bearing and wear out the hub. Too loose an adjustment will chip the bearings at the smaller end.

(2) *Packing the bearings.* Clean the bearings thoroughly (par. 47) and check for wear, chipping or scoring, or corrosion. Dip the bearings in preservative oil, completely drain and then knead the prescribed grease

Figure 86. Unlocking beadlock hinge.

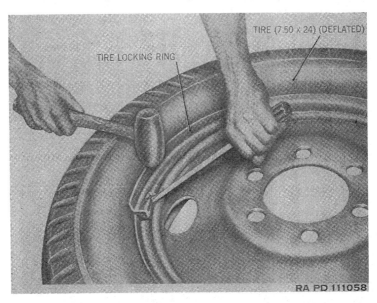

Figure 87. Removing tire locking ring on commercial type wheels.

(par. 42) into the roller cage until it comes out of the other side and around the rollers. Do not pack the hubs full of grease; however, corrosion is deterred if the spindle and the inside of the hub cavity and hub cap are thinly coated with grease or oil. Do not get grease onto the brake bands or brake drum.

Section XX. BRAKE MECHANISM

104. Description and Functioning

Hand brakes of either the clasp (fig. 1) or the plunger type (figs. 2 and 88) are provided for use in parking the howitzer when disengaged from the prime mover. Pulling the brake lever forward, away from the carriage, rotates the splined brake cam shaft forward and actuates an integral eccentric stud to which the brake cam is mounted (inset, fig. 89). This action moves the brake cam out radially, expanding and holding the brake band lining against the brake drum. A ratchet and rack hold the brake lever in the "set" position. When the brake lever is released, the brake band return springs draw the brake band away from the drum to the "released" position.

105. Removal of Brake Mechanism

a. To remove the brake lever, either type, remove the bolt and nut and the cap screw holding the ratchet rack to the brake rack plate. Remove the cotter pin and loosen the castle nut locking the lever to the brake cam shaft (fig. 88). The brake lever together with the rack can then be pulled off the splined brake cam shaft.

b. Remove the hub and brake drum (par. 99) to provide access to the brake band lining and the internal parts of the brake mechanism.

106. Installation of Brake Mechanism

a. Install the brake lever and adjust for proper functioning (par. 109).

b. Tighten the castle nut which locks the brake lever to the brake cam shaft and insert and spread the cotter pin.

c. Install the spacers between the rack and the rack plate and secure the rack to the plate with the bolt and nut and the cap screw.

d. Install hub and brake drum if removed (par. 100).

107. Disassembly of the Plunger Type Brake Lever Assembly

a. Remove the brake lever (par. 105).

b. Depress the plunger cap to release the pressure on the rack catch

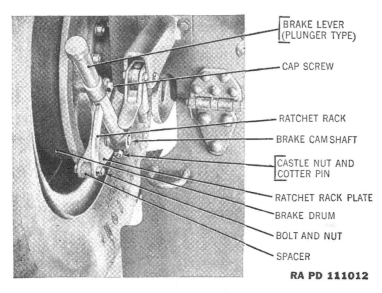

BRAKE LEVER
(PLUNGER TYPE)

CAP SCREW

RATCHET RACK

BRAKE CAMSHAFT

CASTLE NUT AND
COTTER PIN

RATCHET RACK PLATE

BRAKE DRUM

BOLT AND NUT

SPACER

RA PD 111012

Figure 88. Brake mechanism with plunger type brake lever.

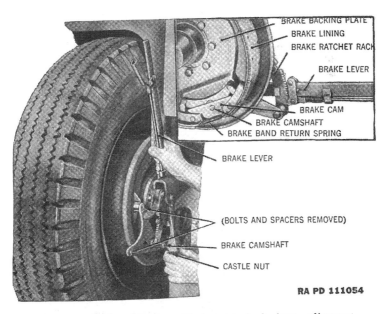

BRAKE BACKING PLATE

BRAKE LINING

BRAKE RATCHET RACK

BRAKE LEVER

BRAKE CAM

BRAKE CAMSHAFT

BRAKE BAND RETURN SPRING

BRAKE LEVER

(BOLTS AND SPACERS REMOVED)

BRAKE CAMSHAFT

CASTLE NUT

RA PD 111054

Figure 89. Brake mechanism with clasp type brake lever—adjustment.

119

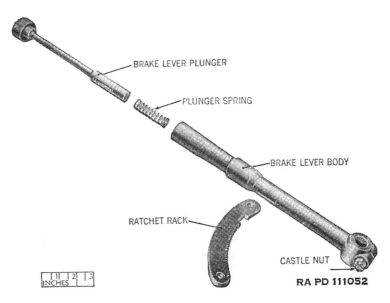

Figure 90. Plunger type brake lever—partially disassembled for maintenance.

and withdraw the rack. Pull the cap and plunger out of the lever and shake out the spring. Figure 90 shows these parts disassembled.

108. Assembly of the Plunger Type Brake Lever Assembly

a. Drop the spring into its recess in the body of the brake lever. Follow with the plunger, making sure that the catch side is back toward the carriage. Depress the plunger cap and insert the rack so that its teeth will be properly engaged by the plunger catch.

b. Adjust the brake (par. 109) and install the brake lever and rack (par. 106).

109. Adjustment of the Brake Mechanism

a. The hand brake lever is properly adjusted when it can be pulled only about halfway forward on the rack before the brake is completely set. This adjustment is obtained by a trial-and-error method. Mark a sharp chalk witness line across the center line of the brake cam shaft and onto the brake lever. Then remove the lever (par. 105) and rotate the lever in relation to the shaft an amount approximately equal to the distance required to secure proper adjustment (fig. 88). Repeat this process until adjustment is obtained. Reinstall the brake lever and rack (par. 106).

b. If proper adjustment cannot be secured on the splined cam shaft, remove the hub (par. 99) and inspect the brake lining for excessive wear. Notify ordnance maintenance personnel for replacement of damaged parts or worn lining.

110. Maintenance of the Brake Mechanism

a. Keep the brakes properly adjusted at all times.

b. Water and mud can work past the ratchet rack and into the body of the plunger type brake lever, thereby fouling the spring, corroding the parts, and causing the ratchet catch to malfunction. This type lever will be disassembled (par. 107), cleaned (par. 47), and oiled (par. 42) at intervals prescribed in paragraph 53. The clasp type lever does not require disassembly for this servicing.

c. Straighten bent or misshapen parts and repaint where necessary (par. 52).

d. At the prescribed service interval for wheel bearings (par. 54) any accumulation of rust, mud, or foreign matter will be cleaned from the internal parts of the brake mechanism. For operation in muddy and rainy areas this interval will be reduced as is necessary for proper functioning (par. 29). Apply a light coating of oil or grease to working parts, but do not allow any lubricant to get onto the braking areas of the drum or brake band linings.

e. The brake cam shaft bearings are serviced periodically and brake band linings are replaced by ordnance maintenance personnel.

PART FOUR

AUXILIARY EQUIPMENT

Section XXI. GENERAL

111. Scope

Part four contains information for the guidance of the personnel responsible for the operation of this equipment. It contains only the information necessary to the using personnel to properly identify, operate, and protect the ammunition, fire control equipment, and subcaliber equipment while being used or transported with the main equipment. Further information, if required, is contained in TMs listed in appendix II.

Section XXII. AMMUNITION

112. General

Ammunition for the 105-mm howitzer M2A1 is issued in the form of semifixed complete rounds. A complete round of semifixed ammunition consists of a cartridge case, a primer, a propelling charge, and a projectile (shell). It is characterized by the free fit of the projectile in the cartridge case. Thus, the propelling charge which is divided into increments for zone firing, may be adjusted in the field just prior to firing. Rounds of new manufacture may be assembled with propelling charges which differ in shape of powder bags; these charges are called "dual-gran." The complete round is loaded into the weapon as a unit.

113. Firing Tables

a. Firing data for the 105-mm howitzer M2A1 are provided in tabular form in FT 105–H–3 and changes thereto. Firing data for the 37-mm subcaliber gun, M13 are prescribed in tabular form in FT 37–BJ–2. Data for use in calculation in preparation and conduct of fire is also provided in graphical form by graphical firing table, GFT M23.

b. FM 21–6 contains a list of firing tables in use. Graphical firing tables are listed in ORD 7, SNL F–237.

114. Classification

Dependent upon the type of projectile, ammunition for the 105-mm howitzer M2A1 is classified as high-explosive (HE), high-explosive antitank (HEAT), chemical (gas or smoke), practice (target-practice), blank, or drill (dummy).

  a. High-explosive shell are comparatively thin-walled projectiles containing a high-explosive bursting charge. For the 105-mm howitzer M2A1 two types of high-explosive shell are provided—

  (1) An HE shell intended primarily for fragmentation and blast effect against personnel.

  (2) An HE, AT shell especially designed for piercing armored target.

  b. Chemical shell contain a chemical filler for producing either a toxic or irritating physiological effect, a screening smoke, an incendiary action, or any combination of these. There are two types of chemical shell: the burster type and the base-ejection type. The burster type contains a centrally located tube of high explosive which ruptures the case and disperses the chemical filler. The base-ejection shell contain a small charge of black powder to provide for the ejection of the shell contents at the time of functioning. Two types of chemical filler are used in chemical base-ejection shell for 105-mm howitzer M2A1:

  (1) HC chemical mixture which produces a white smoke for spotting effect.

  (2) Smoke-producing mixtures of different colors for target identification.

  c. Leaflet or similar filler for propaganda purposes is provided for use with the M84 base-ejection shell for 105-mm howitzer M2A1.

  d. Practice projectiles for 105-mm howitzer M2A1 are inert projectiles of approximately the same size, shape, and weight as the service projectiles which they simulate.

  e. Blank ammunition contains no projectiles.

  f. Drill ammunition, which is completely inert, is intended for practice in loading and handling.

115. Identification

  a. GENERAL. The various rounds are marked as follows (see figs. 91 to 97):

  (1) *On the projectile (stenciled):*
      Weight-zone marking (on HE and chemical shell).
      Caliber and type of weapon in which fired.
      Kind of filler, as "TNT," "WP SMOKE," etc.
      Model of projectile.

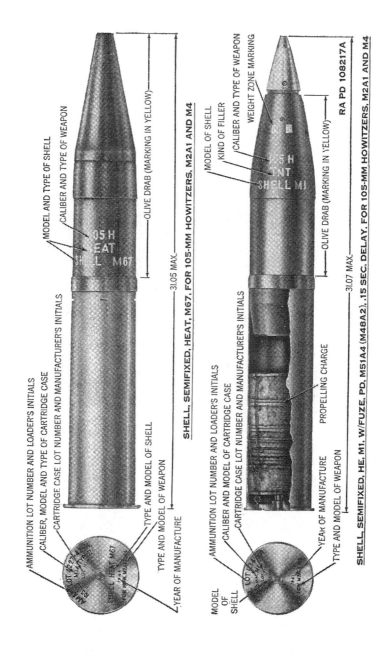

AMMUNITION LOT NUMBER AND LOADER'S INITIALS
CALIBER, MODEL AND TYPE OF CARTRIDGE CASE
CARTRIDGE CASE LOT NUMBER AND MANUFACTURER'S INITIALS
MODEL AND TYPE OF SHELL
CALIBER AND TYPE OF WEAPON
OLIVE DRAB (MARKING IN YELLOW)
TYPE AND MODEL OF WEAPON
TYPE AND MODEL OF SHELL
YEAR OF MANUFACTURE
31.05 MAX.

05 H
EAT
SH LL M67

MODEL
OF
SHELL

SHELL, SEMIFIXED, HEAT, M67, FOR 105-MM HOWITZERS, M2A1 AND M4

AMMUNITION LOT NUMBER AND LOADER'S INITIALS
CALIBER AND MODEL OF CARTRIDGE CASE
CARTRIDGE CASE LOT NUMBER AND MANUFACTURER'S INITIALS
MODEL OF SHELL
KIND OF FILLER
CALIBER AND TYPE OF WEAPON
WEIGHT ZONE MARKING
OLIVE DRAB (MARKING IN YELLOW)
PROPELLING CHARGE
YEAR OF MANUFACTURE
TYPE AND MODEL OF WEAPON
31.07 MAX.

5 H
TNT
SHELL M1

RA PD 108217A

SHELL, SEMIFIXED, HE, M1, W/FUZE, PD, M51A4 (M48A2), .15 SEC. DELAY, FOR 105-MM HOWITZERS, M2A1 AND M4

Figure 91. Ammunition for 105-mm howitzers M2A1 and M4.

124

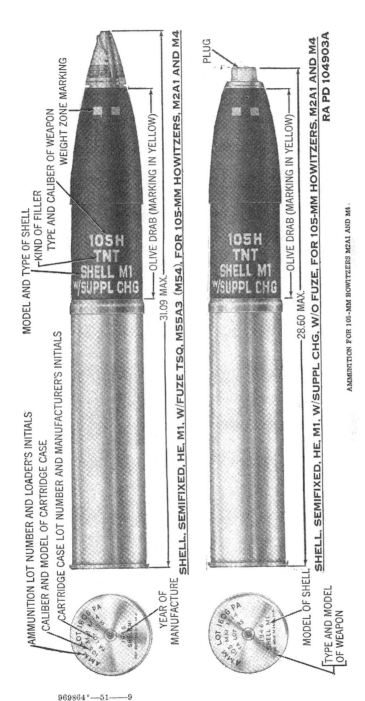

MODEL AND TYPE OF SHELL
KIND OF FILLER
TYPE AND CALIBER OF WEAPON
WEIGHT ZONE MARKING

105 H
TNT
SHELL M1
W/SUPPL CHG

OLIVE DRAB (MARKING IN YELLOW). FOR 105-MM HOWITZERS, M2A1 AND M4

31.09 MAX.

SHELL, SEMIFIXED, HE, M1, W/FUZE TSQ, M55A3 (M54). FOR 105-MM HOWITZERS, M2A1 AND M4

AMMUNITION LOT NUMBER AND LOADER'S INITIALS
CALIBER AND MODEL OF CARTRIDGE CASE
CARTRIDGE CASE LOT NUMBER AND MANUFACTURER'S INITIALS

YEAR OF
MANUFACTURE

PLUG

105 H
TNT
SHELL M1
W/SUPPL CHG

OLIVE DRAB (MARKING IN YELLOW). FOR 105-MM HOWITZERS, M2A1 AND M4
RA PD 104903A

28.60 MAX.

SHELL, SEMIFIXED, HE, M1, W/SUPPL CHG, W/O FUZE, FOR 105-MM HOWITZERS, M2A1 AND M4

AMMUNITION FOR 105-MM HOWITZERS M2A1 AND M4

MODEL OF SHELL

TYPE AND MODEL
OF WEAPON

Figure 92. Ammunition for 105-mm howitzers M2A1 and M4.

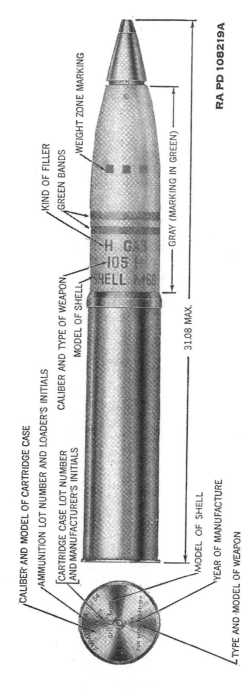

CALIBER AND MODEL OF CARTRIDGE CASE

AMMUNITION LOT NUMBER AND LOADER'S INITIALS

CALIBER AND TYPE OF WEAPON

KIND OF FILLER

GREEN BANDS

WEIGHT ZONE MARKING

MODEL OF SHELL

CARTRIDGE CASE LOT NUMBER AND MANUFACTURER'S INITIALS

GRAY (MARKING IN GREEN)

31.08 MAX.

MODEL OF SHELL

YEAR OF MANUFACTURE

TYPE AND MODEL OF WEAPON

RA PD 108219A

Figure 93. Shell, semifixed, gas, persistent, H, M60, w/fuze, PD, M57, for 105-mm howitzers M2A1 and M4.

126

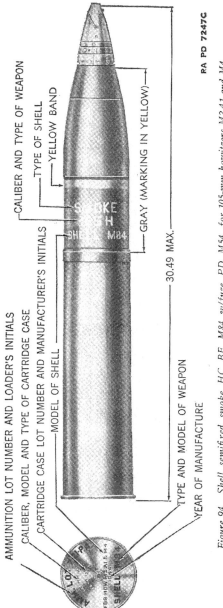

AMMUNITION LOT NUMBER AND LOADER'S INITIALS

CALIBER, MODEL AND TYPE OF CARTRIDGE CASE

CARTRIDGE CASE LOT NUMBER AND MANUFACTURER'S INITIALS

MODEL OF SHELL

CALIBER AND TYPE OF WEAPON

TYPE OF SHELL

YELLOW BAND

GRAY (MARKING IN YELLOW)

30.49 MAX.

TYPE AND MODEL OF WEAPON

YEAR OF MANUFACTURE

RA PD 7247C

Figure 94. Shell, semifixed, smoke, HC, BE, M84, w/fuze, PD, M54, for 105-mm howitzers M2A1 and M4.

127

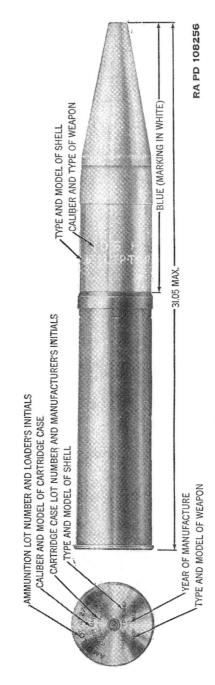

AMMUNITION LOT NUMBER AND LOADER'S INITIALS

CALIBER AND MODEL OF CARTRIDGE CASE

CARTRIDGE CASE LOT NUMBER AND MANUFACTURER'S INITIALS

TYPE AND MODEL OF SHELL

YEAR OF MANUFACTURE

TYPE AND MODEL OF WEAPON

TYPE AND MODEL OF SHELL

CALIBER AND TYPE OF WEAPON

BLUE (MARKING IN WHITE)

31.05 MAX.

RA PD 108256

Figure 95. Shell, semifixed, TP–T, M67, for 105-mm howitzers M2A1 and M4.

128

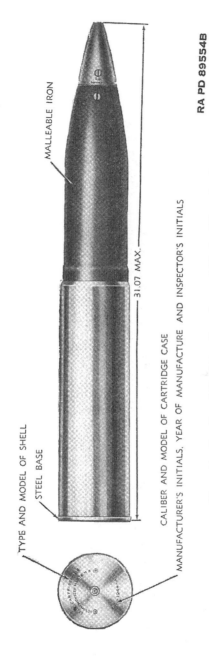

MALLEABLE IRON

TYPE AND MODEL OF SHELL

STEEL BASE

31.07 MAX.

CALIBER AND MODEL OF CARTRIDGE CASE

MANUFACTURER'S INITIALS, YEAR OF MANUFACTURE AND INSPECTOR'S INITIALS

Figure 96. Cartridge, drill, M14, w/fuze, dummy, M59, for 105-mm howitzers.

RA PD 89554B

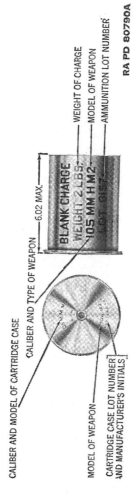

CALIBER AND MODEL OF CARTRIDGE CASE

CALIBER AND TYPE OF WEAPON

MODEL OF WEAPON

CARTRIDGE CASE LOT NUMBER
AND MANUFACTURER'S INITIALS

BLANK CHARGE
WEIGHT 2 LBS.
105 MM H M2
LOT 8187

6.02 MAX.

WEIGHT OF CHARGE
MODEL OF WEAPON
AMMUNITION LOT NUMBER

RA PD 80790A

Figure 97. Ammunition, blank, for 105-mm howitzers.

(2) *On the fuze (stamped in the metal):*
   Type and model of fuze. (Since M48A2 and M51A4 fuzes may
   have an 0.05 or 0.15 second delay element, the length of delay
   is stamped on the fuze following the model number. In addition to the stamping on the fuze, nondelay CP M78 fuzes
   have the tip painted white; tips of the M78 fuzes with 0.025
   second delay are not painted.)

(3) *On the base of the cartridge case:*
   Ammunition lot number (stenciled).
   Model of projectile (stenciled).
   Type and model of weapon (stenciled).
   Caliber and model of cartridge case (stamped).
   Cartridge case lot number, including manufacturer's initials
   (stamped).
   Year of manufacture (stamped).

b. AMMUNITION LOT NUMBER. A lot number is assigned to all ammunition at the time of manufacture. It is marked on every loaded complete
round and on all packing containers. It is required for all purposes of
record, including reports on condition, functioning, and accidents in
which the ammunition is involved.

c. WEIGHT-ZONE MARKINGS. It is not practicable to manufacture
projectiles within the narrow weight limits required for accuracy of fire.
Therefore they are grouped into weight zones and appropriate ballistic
corrections are given in the firing tables for the variation in weight.
The weight zone of the projectile is indicated thereon by squares of the
same color as the original markings. There are one, two, three, four,
or more such squares, dependent upon the weight of the projectiles. For
the 105-mm howitzers, two squares denote "normal" or "standard"
weight. Base-ejection and high-explosive-antitank shells do not have
weight-zone markings.

d. PAINTING. Artillery projectiles are painted primarily to prevent
rust and secondarily to provide, by the color, a means of identification
as to type. The color scheme is as follows:

(1) High-explosive—olive drab, markings in yellow.

(2) Casualty-gas—gray, markings in green; two green bands denoting a persistent casualty gas.

(3) Harassing gas—gray, markings in red; two red bands denoting
a persistent harassing gas.

(4) Smoke—gray, markings in yellow; one yellow band to denote
a smoke-producing filler.

(5) Practice—blue, markings in white.

(6) Drill or dummy (inert)—unpainted, because of cadmium-plated
and bronze parts. Dummy or drill ammunition made of ferrous metals
is normally painted black, markings being in white.

116. Care, Handling, and Preservation

a. Complete rounds and ammunition components are packed to withstand conditions usually found in the field. Care must be taken to keep packing boxes from becoming broken or damaged. All broken boxes must be repaired immediately and all markings transferred accurately to the new parts of the box. Since explosives are adversely affected by moisture and high temperature, due consideration should be given to the following:

(1) Do not break the moisture-resistant seal until the ammunition is to be used.

(2) Protect ammunition, particularly fuzes, from sources of high temperature, including direct rays of the sun. More uniform firing is obtained if all rounds are at the same temperature.

b. Do not attempt to disassemble any fuze.

c. Before loading the complete round into the weapon, each of the components should be free of foreign matter, sand, mud, moisture, grease, etc.

d. Do not remove protection or safety devices from fuzes until just before use.

e. Explosive ammunition or components containing explosives must be handled with appropriate care at all times. The explosive elements in primers and fuzes are particularly sensitive to undue shock and high temperature.

f. Rounds prepared for firing but not fired will be returned to their original condition and packings and appropriately marked. Such rounds will be used first in subsequent firings in order that stocks of opened packings may be kept at a minimum.

g. Do not handle duds. Because their fuzes are armed, and hence extremely dangerous, duds will not be moved or turned, but will be destroyed in place in accordance with TM 9–1900. Unlike other fuzes, duds containing VT fuzes may be considered safe for handling 24 hours after the firing of the projectile, but they should be handled with care since they contain an unignited powder train and booster charge.

117. Storage and Handling of VT Fuzes

a. Precautions applying to other packed ammunition also apply to VT fuzes. In addition, storage temperature limits should be held within −20° and +130° F. Storage outside these limits for any length of time will result in permanent damage. The direct rays of the sun on VT fuze containers may cause the temperature inside the container to exceed 130° F. and must be avoided.

b. VT fuzes must be protected against dampness. Although the fuzes are nearly waterproof, any exposure to dampness may increase the number of duds. Contact with rain or immersion in water will hasten deterioration. Particularly in tropical climates, the storage time of unpacked fuzes should be kept to a minimum. In other climates, fuzes can be safely used after two months storage outside of their packing containers but should be stored in the original sealed metal containers insofar as practicable.

c. VT fuzes will withstand normal handling without danger of detonation or damage when in their original packing containers or when assembled to projectiles. However, care should be taken not to strike or drop fuzes or fuzed rounds as these actions may increase the number of duds. A drop of four feet in certain positions may cause a dud. Rough handling may not decrease fuze safety but may increase the number of duds.

d. VT-fuzed ammunition may be safely transported short distances with normal care in handling. However, when such ammunition is to be transported considerable distances it may be advisable to remove the fuze from the shell and return the fuze to its original marked container. The supplementary charge and original fuze or closing plug (with gasket and spacer) should be reassembled to the shell, making certain that the supplementary charge is inserted properly (felt-pad end innermost, lifting strap outermost).

e. Fuzes and supplementary charges which have been removed from the shell will be packed in the containers from which VT fuzes have been removed. The containers should be properly marked and returned to ordnance personnel for disposition.

f. When rounds on which fuzes have been changed are returned to their containers, care must be taken to change markings on the containers and boxes to conform with the change in ammunition.

g. Rounds fuzed with VT fuzes must be specially padded when returned to their fiber containers. The U-shaped support which engages the wrench slots of time or impact fuzes will not fit the slots in VT fuzes and must, therefore, be omitted. The play that results is taken up by placing extra corrugated board pads under the base end of the projectile before closing the container.

118. Authorized Rounds

a. Ammunition authorized for use in the 105-mm howitzer M2A1 is listed in Table III. Standard nomenclature, which completely identifies the ammunition, is used in the listing.

b. Some high-explosive shell M1 are loaded to provide a deep fuze cavity so that these shell can be used with either VT fuzes, or upon insertion of a supplementary bursting charge of TNT, with other standard fuzes. Rounds with this shell are not shipped completely assembled with the supplementary charge and standard impact, or time-and-impact, fuze. Such rounds are marked "W/SUPPL CHG" on the shell. Early shipments of deep cavity shell were assembled with or without supplementary charge but without fuze, the fuze hole being closed by a nose plug. Such shell are marked "FOR VT FUZE."

c. For target and battery identification purposes, the HC smoke shell M84 may be converted by replacing the HC-filled canisters with canisters of the M2 type which contain colored-smoke mixtures. CANISTER, M2, is provided in four colors of red, green, violet, and yellow, for such replacement in 105-mm howitzer smoke shell. Shell assembled prior to issue with colored-smoke-filled canisters are marked to indicate the color.

d. FUZE, CP, M78 (T105), 0.025-sec delay, w/booster, M25 (T1E1) and FUZE, CP, M78 (T105), nondelay, w/booster, M25 (T1E1) (fig. 99) are authorized for use with high-explosive shell for the 105-mm howitzers. These concrete-piercing fuzes, like the VT fuzes, are issued separately for assembly in the field.

e. For future manufacture of M60 chemical shell, the M51A4 fuze with 0.15-second delay is authorized as standard and the M57 fuze is limited standard.

f. FLASH REDUCER, M4, is provided for reducing flash during night firing for 105-mm howitzers, M2A1 and M4.

Table III. Authorized Ammunition for 105-mm Howitzer, M2A1

Standard nomenclature	Complete round		Projectile			Action of fuze
	Weight (lb)	Length (in.)	Weight as fired (lb)	Type of filler	Weight (lb)	
Service Ammunition[1]						
SHELL, semifixed, HE, M1, dual-gran, w/fuze, PD, M51A4, 0.15 sec, delay, for 105-mm howitzers, M2A7 and M4.	42.00	31.07	33.00	TNT	4.84	SQ or 0.15 sec delay.
SHELL, semifixed, HE, M1, w/fuze, PD, M51A4 (M48A2), 0.15 sec delay, for 105-mm howitzers, M2A1 and M4.	42.06	31.07	33.00	TNT	4.84	SQ or 0.15 sec delay.
SHELL, semifixed, HE, M1, w/fuze, PD, M48A1, 0.15 sec delay, for 105-mm howitzers, M2A1 and M4.	42.06	31.07	33.00	TNT	4.84	SQ or 0.15 sec delay.

134

Table III. Authorized Ammunition for 105-mm Howitzer, M2A1—Continued

Standard nomenclature	Complete round		Projectile			Action of fuze
	Weight (lb)	Length (in.)	Weight as fired (lb)	Type of filler	Weight (lb)	
Service Ammunition[1]—Continued						
SHELL, semifixed, HE, M1, w/fuze, PD, M48, 0.05 sec delay, for 105-mm howitzers, M2A1 and M4.	42.07	31.07	33.00	TNT	4.84	SQ or 0.05 sec delay.
SHELL, semifixed, HE, M1, dual-gran, w/fuze, TSQ,, M55A3, for 105-mm howitzers, M2A1 and M4.	42.00	31.09	33.00	TNT	4.84	Time or SQ.
SHELL, semifixed, HE, M1, w/fuze, TSQ, M55A3 (M54), for 105-mm howitzers, M2A1 and M4.	42.06	31.09	33.00	TNT	4.84	Time or SQ.
SHELL, semifixed, HE, M1, w/suppl chg and fuze, PD, M51A4 (M48A2), 0.15 sec delay, for 105-mm howitzers, M2A1 and M4.	41.84	31.07	32.89[3]	TNT[4]	4.37[4]	SQ or 0.15 sec delay.
SHELL, semifixed, HE, M1, w/suppl chg and fuze, PD, M48A1, 0.15 sec delay, for 105-mm howitzers, M2A1 and M4.	41.84	31.07	32.89[3]	TNT[4]	4.37[4]	SQ or 0.15 sec delay.
SHELL, semifixed, HE, M1, w/suppl chg and fuze, TSQ, M55A3 (M54), for 105-mm howitzers, M2A1 and M4.	41.84	31.09	32.89[3]	TNT[4]	4.37[4]	Time or SQ.
SHELL, semifixed, HE, M1, w/suppl chg w/o fuze, for 105-mm howitzers, M2A1 and M4.	40.06	28.60[5]	33.12[6]	TNT[4]	4.37[4][7]	(VT; SQ or delay, or time or SQ.)
SHELL, semifixed, HE, M1, w/o fuze, for VT fuze, for 105-mm howitzers, M2A1 and M4.	39.76	28.60[5]	33.12[6]	TNT[4]	4.37[4][7]	(VT)

*Table III. Authorized Ammunition for 105-mm Howitzer, M2A1—*Continued

Standard nomenclature	Complete round		Projectile			Action of fuze
	Weight (lb)	Length (in.)	Weight as fired (lb)	Type of filler	Weight (lb)	
Service Ammunition[1]—Continued						
SHELL, semifixed, HE, AT, M67, for 105-mm howitzers, M2A1 and M4[2].	36.89	31.05	29.12	Pentolite	2.93	Nondelay.
SHELL, semifixed, smoke, WP, M60, w/ fuze, PD, M57, for 105-mm howitzers, M2A1 and M4.	43.42	31.08	34.35	WP	4.06	SQ.
SHELL, semifixed, smoke, FS, M60, w/ fuze, PD, M57, for 105-mm howitzers, M2A1 and M4.	43.93	31.08	34.86	FS	4.61	SQ.
SHELL, semifixed, gas, persistent, H, M60, dual-gran, w/fuze, PD, M57, for 105-mm howitzers M2A1 and M4.	42.92	31.08	33.42	H	3.17	SQ.
SHELL, semifixed, gas, persistent, H, M60, w/fuze, PD, M57, for 105-mm howitzers, M2A1 and M4.	42.49	31.08	33.42	H	3.17	SQ.
SHELL, semifixed, gas, persistent, CNS, M60, w/fuze, PD, M57, for 105-mm howitzers, M2A1 and M4.	42.68	31.08	33.61	CNS	3.32	SQ.
SHELL, semifixed, smoke, HC, BE, M84, w/fuze, TSQ, M54, for 105-mm howitzers, M2A1 and M4.	41.94	30.49	32.87	HC mixture	7.50	Time or SQ.
SHELL, semifixed, smoke, yellow, BE, M84, w/fuze, TSQ, M54, for 105-mm howitzers, M2A1 and M4.	39.37	30.49	30.28	Yellow smoke mixture	4.92	Time or SQ.

Table III. Authorized Ammunition for 105-mm Howitzer, M2A1—Continued

Standard nomenclature	Complete round		Projectile			Action of fuze
	Weight (lb)	Length (in.)	Weight as fired (lb)	Type of filler	Weight (lb)	

Service Ammunition[1]—Continued

Standard nomenclature	Complete round		Projectile			Action of fuze
SHELL, semifixed, smoke, red, BE, M84, w/fuze, TSQ, M54, for 105-mm howitzers, M2A1 and M4.	39.77	30.49	30.68	Red smoke mixture	5.32	Time or SQ.
SHELL, semifixed, smoke, violet, BE, M84, w/fuze, TSQ, M54, for 105-mm howitzers, M2A1 and M4.	39.57	30.49	30.48	Violet smoke mixture	5.12	Time or SQ.
SHELL, semifixed, smoke, green, BE, M84, w/fuze, TSQ, M54, for 105-mm howitzers, M2A1 and M4.	39.57	30.49	30.48	Green smoke mixture	5.12	Time or SQ.
SHELL, semifixed, propaganda, BE, M84, w/fuze, TSQ, M54, for 105-mm howitzers, M2A1 and M4.	34.44	30.49	25.37	Black powder[3]	0.14[8]	Time or SQ.

Practice Ammunition

Standard nomenclature	Complete round		Projectile			Action of fuze
SHELL, semifixed, TP-T, M67, for 105-mm howitzers, M2A1 and M4[2].	36.96	31.05	29.19	Inert material	3.79	None[9].
SHELL, semifixed, empty, M1, w/fuze, PD, M48, inert, for 105-mm howitzers, M2A1 and M4.	37.22	31.07	33.00	Inert material	4.84	None.

Blank Ammunition[1]

Standard nomenclature	Complete round		Projectile			Action of fuze
AMMUNITION, blank, for 105-mm howitzers.	6.24	6.02	....	....	....	....

Table III. Authorized Ammunition for 105-mm Howitzer, M2A1—Continued

Standard nomenclature	Complete round		Projectile			Action of fuze
	Weight (lb)	Length (in.)	Weight as fired (lb)	Type of filler	Weight (lb)	
Drill Ammunition						
CARTRIDGE, drill, M14, w/fuze, dummy, M59, for 105-mm howitzers[10].	41.43	31.07	....	....	....	....
CARTRIDGE, drill, M14, w/fuze, TSQ, M54, inert, for 105-mm howitzers[10].	41.42	31.07	....	....	....	....

AT—antitank.
BE—base ejection.
HE—high-explosive.
PD—point detonating.
sec—second.
SQ—superquick.
suppl—supplementary.
TP-T—target-practice, with tracer.
TSQ—time and superquick

[1] Steel case rounds are substitute standard.
[2] The propelling charge is fixed, that is, not adjustable.
[3] As fired with fuze assembled as shipped.
[4] Does not include supplementary charge assembly, weight 0.295 pound.
[5] With closing plug 75–14–575A; with closing plug 75–14–575C, length is 29.73 inches.
[6] As fired with VT fuzes; weight with M48, M54, M51, M55 fuzes, 32.89 pounds.
[7] May be fired with impact, time, CP, or, VT fuzes.
[8] Black powder ejection charge.
[9] Earlier projectiles were fitted with a dummy base fuze; present manufacture has a base plug.
[10] Body may be made of bronze or malleable iron.

119. Interchangeability

a. In addition to the rounds listed in Table IV, corresponding high-explosive rounds M1 for the 105-mm howitzer M3 also are authorized for use in the howitzers M2A1, as substitute standard. The rounds differ only in that those for the howitzer M3 have a five-section propelling charge of quick-burning powder. While increments cannot be interchanged, an M3 howitzer round can be fired in the howitzer M2A1 with any or all of its five charges. Firing data is obtained by applying appropriate corrections in FT 105–H–3 and in Changes No. 8 thereto.

b. HE, AT rounds for the M3 and for the M2A1 and M4 howitzers are *not* interchangeable between howitzers. They will be fired only in the weapons for which designed in order that maximum penetration may be obtained and dangerous accidents avoided.

120. Preparation for Firing

a. GENERAL. Rounds for the 105-mm howitzer M2A1 require preparation of shell, propelling charge, and fuze as described below, with the exception of the HEAT round M67, which has a fixed (nonadjustable) propelling charge and single-action base fuze and is, therefore, ready for firing upon removing packing material.

Note. Upon removing a round from its fiber container, withdraw the U-shaped packing stop, when this stop is present, from the fuze wrench slots in the fuze. This stop is used with the packed projectile to prevent the fuze from touching the separating partitions in the center of the fiber container or from touching the ends of containers. Serious damage may result if this stop is not removed before firing.

b. SHELL. (1) Rounds to be fired with their original fuzes require only the adjustment of fuzes as described in paragraph 121.

(2) To prepare rounds for firing with CP fuzes (fig. 99), proceed as follows:

(*a*) Place round to be refuzed on its side. Protect the primer in the base of the cartridge case from being struck or damaged and the cartridge case from being dented.

(*b*) The booster set-screw, when present, must be loosened or removed. Loosen the set-screw with a screw driver which fully fits the screw.

(*c*) Insert fuze wrench M16 (fig. 39), provided for the purpose, in the fuze slots and strike the wrench sharply in a counterclockwise direction to loosen the fuze from the shell, taking care to avoid striking any part of the fuze. Remove the fuze. If the booster comes out with the fuze, proceed to step (*e*) below.

(*d*) Remove the booster, using the booster end of the fuze wrench M16.

(*e*) Examine the fuze threads in the shell and the threads on the booster M25 and fuze M78 to insure that they are in good condition. Do not use components with damaged threads.

(*f*) Remove the safety pin from the booster M25 and screw the booster into the booster cavity of the shell. Tighten the booster firmly with the booster end of the fuze wrench. Boosters which are issued without safety pins should not be used.

(*g*) Screw the fuze M78 into the fuze cavity and tighten it securely. Make sure the fuze shoulder seats firmly against the nose of the shell. There should be no space between the fuze shoulder and the shell. Do not stake the fuze to the shell.

(*h*) If possible, the booster screw, when present, should be screwed back into the shell.

(3) Deep-cavity shell issued with standard impact and time-and-impact fuzes are prepared for firing with VT fuzes as follows:

(*a*) Remove assembled fuze, using a fuze wrench and turning, with fuze up, in counterclockwise direction. (As the fuze is staked to the shell, it may be necessary to strike the handle of the wrench sharply to loosen it).

Note. Do not remove the wax plug from the set-screw hole in the front of the shell; all shell of new manufacture will omit this hole.

(*b*) Remove the supplementary charge by means of its cloth tape loop.

(*c*) Inspect the cavity for damage. Remove any loose material from the cavity. If the HE filler around the cavity appears to have been broken, reject the shell. If any HE is found adhering to the threaded portion of the shell throat, remove it with a pointed instrument made of wood or a nonferrous metal.

(*d*) Screw in the VT fuze by hand. If binding occurs, inspect the fuze cavity, and threads of both fuze and shell. Reject whichever is at fault.

(*e*) Tighten the fuze to the shell with the special fuze wrench M18 (fig. 39), issued with boxes of VT fuzes. Use only such force as can be applied by hand to the fuze wrench handle. If the fuze cannot be tightened to form a good seat between the shell and fuze, reject the component at fault. *Do not hammer on the wrench or use an extension handle. Do not stake the fuze to the shell under any circumstances.*

(4) To prepare shell marked "FOR VT FUZES" for firing with such fuzes—

(*a*) Remove the closing plug and gasket, and supplementary charge if one is present.

(*b*) Inspect the fuze cavity.

(*c*) Assemble the fuze to the shell as outlined in step (3)(*d*) and (*e*) above.

(5) To prepare shell marked "FOR VT FUZE" for firing with time or impact fuze—

(*a*) Remove the closing plug and gasket.

(*b*) Inspect the fuze cavity.

(*c*) Properly insert supplementary charge (felt-pad end innermost and lifting strap outermost).

(*d*) Assemble the fuze to the shell in the usual way.

c. Propelling Charge. Adjust the propelling charge for the zone to be fired. (An exception is HE, AT, semifixed shell M67 which has a nonadjustable charge.) Proceed as follows: Remove the projectile from the cartridge case, being careful not to damage the lip of the case. If the lip is damaged, the round may jam in the chamber of the howitzer. Withdraw the increments from the cartridge case, and remove and discard those increments numbered higher than the zone to be fired by cutting or breaking the twine between the designated zone and the higher numbered increments. Reassemble the remaining increments (from 1 up to and including the number of the zone to be fired) in the cartridge case in their original numerical order with the number of each increment uppermost. For example, when adjusting the seven-section charge for zone 4, increments 5, 6, and 7 will be removed and the remaining increments 1, 2, 3, and 4 will be reassembled in the cartridge case. The round, insofar as the propelling charge is concerned, is now ready for firing. When firing the full (outer zone) charge, no adjustment is required, the full charge as issued being used.

d. FUZES. Fuzes M62, and M62A1 and M91, single-action, base-detonating types; fuze M57, a single-action point fuze; and VT fuzes, which by their nature are automatic or self-setting types, do not require preparation for firing. Other fuzes used with ammunition described in this section require adjustment of setting as described in paragraph 121.

121. Fuzes

a. GENERAL. A fuze is a mechanical device used with a projectile to explode it at the time and under the circumstances desired.

b. CLASSIFICATION. Fuzes are classified according to their manner of action as "time" or "impact." Time fuzes are either automatic, self-acting types which function on approach to the target ("variable time"), or adjustable types which contain a graduated element in the form of a compressed black powder train or mechanism similar to clockwork, to explode the shell a certain number of seconds after firing. Impact fuzes function when the projectile strikes a resistant object. Impact types are classified according to rapidity of action as superquick, nondelay, and delay. According to their location on the projectile, detonating fuzes are known as point-detonating (PD) or base-detonating (BD).

c. BORESAFE AND NONBORESAFE. A boresafe (detonator-safe) fuze is one in which the explosive train is so interrupted that prior to firing and while the projectile is still in the bore of the cannon, premature action of the bursting charge is prevented should any of the more sensitive elements, primer and/or detonator, malfunction. The fuzes are classified as follows:

Boresafe	*Nonboresafe*
FUZE, PD, M51A4 (M48A2)	FUZE, PD, M57 (Not considered bore-safe because it is used in conjunction with the nonboresafe booster M22 assembled to the shell.)
FUZE, PD, M48A1	
FUZE, PD, M48	
FUZE, PD, M54	
FUZE, PD, M55A3 (M54)	
FUZE, PD, M51A4	
FUZE, CP, M78 (T105), w/booster, M25 (T1E1)	
FUZE, VT, M97 (T80E9)	
FUZE, BD, M62	
FUZE, BD, M62A1	
FUZE, BD, M91	

Caution. Fuzes will not be disassembled. Any attempt to disassemble fuzes in the field is dangerous and is prohibited except under specific directions from the Chief of Ordnance.

d. FUZE, M48, M48A1, M48A2 OR M51A4. (1) *Description.* This fuze (fig. 98), shown as issued fitted to the projectile in figure 91, is a

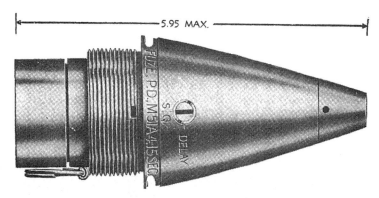

M51A4 PD FUZE .15 SEC DELAY

M55A3 TSQ FUZE   **RA PD 109516**

Figure 98. PD and SQ Fuzes.

combination superquick and delay type. The delay action for fuze M48 is 0.05 second; for fuze M48A1, 0.15 second. Fuze M48A2 is manufactured with 0.05-second or with 0.15-second delay, dependent on the lot. The M51A4 differs from the M48A2 by having a booster assembled to it and handled as a unit with the fuze. On the side of the fuze, near the base, is a slotted "setting sleeve" and two registration lines. One line is parallel to the axis of the fuze and marked "S.Q." (superquick), the other at right angles thereto and marked "DELAY." To set the fuze, the setting sleeve is turned with a screw driver or similar instrument, for example fuze wrench M18, so that the slot is alined with "S.Q." or "DELAY," whichever is required. The setting is made any time before firing, even in the dark, by noting the position of the slot parallel to the fuze axis for "S.Q." at right angles thereto for "DE-LAY." It should be noted that, even though set superquick, this fuze

142

will function with delay action should the superquick action fail to function.

(2) *Preparation for firing.* Prior to firing, it is only necessary to set the fuze as described above, and this only when delay action is required, since, as shipped, the fuze is set superquick.

e. TSQ FUZE M54 OR M55A3. (1) *Description.* This fuze (fig. 98), shown as issued assembled to the projectile in figure 92, is a combination time and superquick type. The M55A3 differs from the M54 by having a booster attached to it. The M54 is used with base-ejection shell whereas the M55A3 is used with high-explosive shell. A safety wire extends through the fuze to secure the time plunger during shipment. The fuze has two actions, time and superquick. The superquick action is always operative and will function on impact unless prior functioning has been caused by time action. Therefore, to set the fuze for superquick action, it is required that the time action be set either at safe (S) or for a time longer than expected time of flight. The time train ring graduated for 25 seconds is similar to that of other powder time train fuzes. To prevent extremely short time action, an internal safety feature prevents the time action from functioning, should the fuze be set for less than 0.4 second. As shipped, the fuze is set safe (S). The fuze is set for time by means of a fuze setter (M14 or M22) (par. 131).

(2) *Preparation for firing.* Prior to firing, with either superquick or time setting, the safety pull wire must be withdrawn from the fuze by pulling the lower end of the wire from the hole and sliding the wire off the end of the fuze. If superquick action is required, the graduated time ring can be left as shipped at safe (S) or set for a time greater than the expected time of flight, using fuze setter M14 or M22 (par. 131).

Note. If the fuze is prepared for firing and is not used, it will be reset safe (S) and the safety pin replaced in its proper position before returning the round to its container.

f. PD FUZE M57. (1) *Description.* This fuze is a superquick type similar in appearance to fuze M48, M48A1, M48A2, or M51A4 except for the marking and the absence of the delay element assembly and setting sleeve.

(2) *Preparation for firing.* This fuze, being a single-action type, requires no setting or other special preparation for firing.

g. FUZE M62, M62A1 OR M91. This fuze is assembled in the base of the antitank projectile (HEAT, Shell M67) and is known as a base-detonating (BD) fuze. It functions upon impact with nondelay action. Because of its location, the fuze is not visible. M91 differs from the M62 and M62A1 by having a tracer in its base end.

h. FUZE CP M78 (T105). This concrete piercing fuze (fig. 99) consists of a solid, hardened steel nose plug with a detonator assembly in

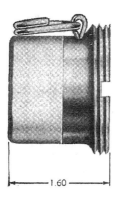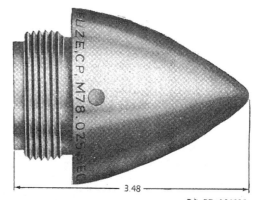

Figure 99. Fuze, CP, M78.

its base equipped to function with 0.025 second delay (for fire for effect), or with nondelay action (for spotting purposes), as marked on the individual fuze. To facilitate identification, the tip of the nondelay fuze is painted in white whereas the tip of the 0.025 second delay fuze is unpainted. Since these are single-action fuzes and there are no external safety devices, no preparation for firing is required.

i. VT Fuze M97 (T80E9). (1) *Description.* The VT fuze M97 is a proximity fuze without impact element, and is provided for use in terrestrial fire with deep-cavity high-explosive shell (without supplementary charge). It is essentially a self-powered radio transmitting and receiving unit. In flight, the armed fuze broadcasts radio waves. When any part of the radio wave front is reflected back from the target, it interacts with the transmitted wave. When the ripple or beat of this interaction reaches a predetermined intensity, it trips a switch which closes an electric circuit and initiates detonation of the fuze explosive train. Boresafety is provided by an arming switch which delays arming of the fuze for approximately five seconds. When armed, the fuze will function on close approach to any object capable of reflecting the transmitted waves.

(2) *Preparation for firing.* Since all functioning within the fuze is automatic, no adjustment in preparation for firing is required. It should be noted that the fuze will function properly at temperatures within 0° and 120° F., and should not be used outside these limits. Also, if the fuzed round is loaded into the chamber of a hot gun and not fired before 30 seconds, the fuze probably will cause either an early burst or a dud.

j. Precautions to be Observed in Firing, M48, M51, M55 and M57 Series Fuzes. If M48, M51, M55, and M57 series fuzes are fired

144

during extremely heavy rainfall, premature functioning may occur which will result in an air burst. The rainfall necessary to cause such malfunctioning is comparable with the **exceedingly** heavy downpours commonly occurring during summer thundershowers. In the case of M48 and M51 series fuzes, occurrences of prematures may be prevented when firing under the conditions described above by setting the fuze for delay action, making the SQ action inoperative. The M55 and M57 fuzes cannot be remedied, however, since the SQ action is always operative.

122. Packing

Rounds for the 105-mm howitzer are packed in the individual fiber container M105, in which the shell is seated in an inverted position (nose within the cartridge case neck). Outer packing consists of two types: For shipment to areas where extraordinary protection against moisture is required, rounds in the fiber container are packed in the gasket-sealed metal container M152 (fig. 100); for other areas, the standard packing consists of a wooden box holding two rounds, each in its individual fiber container (fig. 101). Weights vary somewhat, dependent upon the type and model of rounds. The two-round wooden box weighs approximately 120 pounds with a volume of 1.8 cubic feet; the metal container packing weighs approximately 71 pounds with a volume of 0.8 cubic feet.

123. Subcaliber Ammunition

a. GENERAL. SHELL, fixed, practice, M92, w/fuze, PD, M74, for 37-mm subcaliber guns, M12, M13, M14, M16, and M1916 is standard for use in the 37-mm subcaliber gun for this howitzer. SHELL, fixed, practice, M63 Mod 1, for 37-mm subcaliber guns, M12, M13, M14, M15, M16, and M1916 is substitute standard. The rounds are issued in the form of fuzed rounds of fixed ammunition (fig. 102). Both projectiles have a black powder charge which serves as a spotting charge, but the shell M92 is fitted with a point-detonating fuze whereas the M63 Mod 1 has a base-detonating fuze.

b. PACKING. The M63 Mod 1 rounds are packed 10, 25, or 40 rounds per box. M92 rounds are packed 40 per box. The 40-round box for the M63 Mod 1 round has a volume of 1.2 cubic feet and weighs, when loaded, about 76 pounds. The same packing for M92 rounds has a volume of 1.5 cubic feet and weighs approximately 93 pounds. Volume of the 25-round pack for M63 Mod 1 rounds is 1.3 cu ft; weight is about 72 pounds, while corresponding data for the 10-round pack are 0.7 cubic feet and 46 pounds. Complete packing data are published in ORD 11 SNL R–1.

RA PD 97688A

Figure 100. Metal container M152, for 105-mm howitzer ammunition.

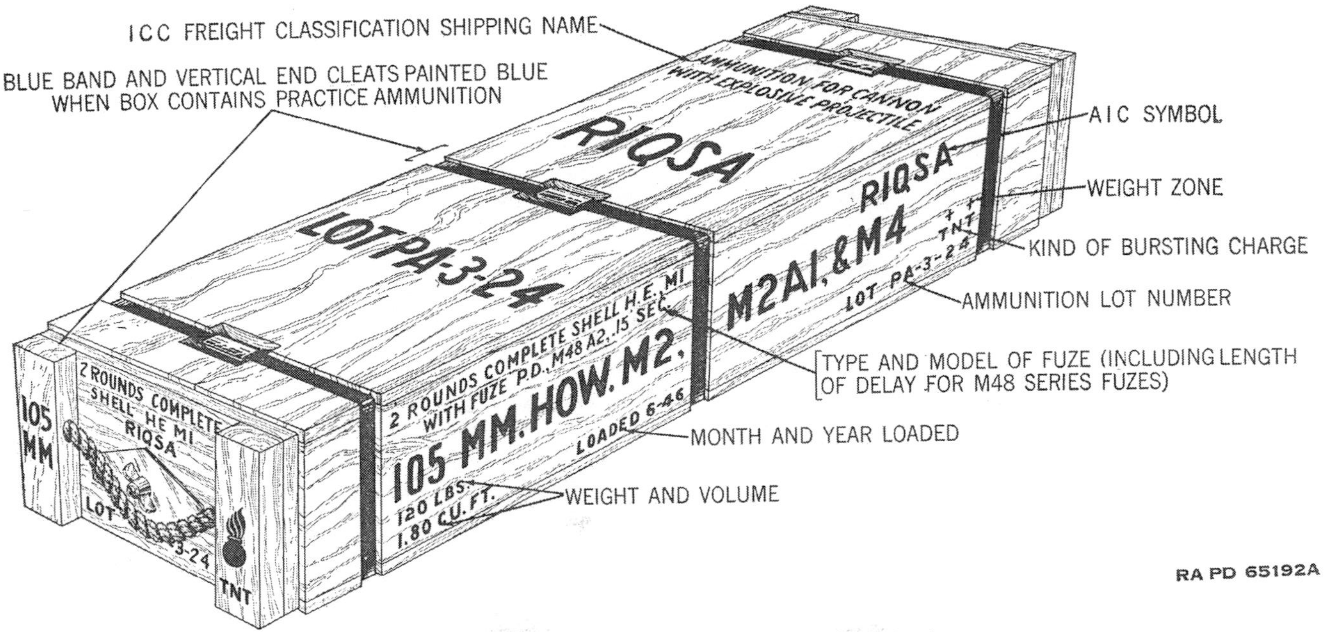
ICC FREIGHT CLASSIFICATION SHIPPING NAME
BLUE BAND AND VERTICAL END CLEATS PAINTED BLUE WHEN BOX CONTAINS PRACTICE AMMUNITION
AIC SYMBOL
WEIGHT ZONE
KIND OF BURSTING CHARGE
AMMUNITION LOT NUMBER
TYPE AND MODEL OF FUZE (INCLUDING LENGTH OF DELAY FOR M48 SERIES FUZES)
MONTH AND YEAR LOADED
WEIGHT AND VOLUME
AMMUNITION FOR CANNON WITH EXPLOSIVE PROJECTILE
RIQSA
LOT PA-3-24
RIQSA
M2A1, & M4
TNT
LOT PA-3-24
2 ROUNDS COMPLETE SHELL H.E. M1
WITH FUZE P.D. M48 A2, .15 SEC.
105 MM. HOW. M2,
LOADED 6-46
120 LBS.
1.80 CU. FT.
105 MM
2 ROUNDS COMPLETE
SHELL HE M1
RIQSA
LOT 3-24
TNT
RA PD 65192A

Figure 101. Long 2-round box, for 105-mm howitzer ammunition.

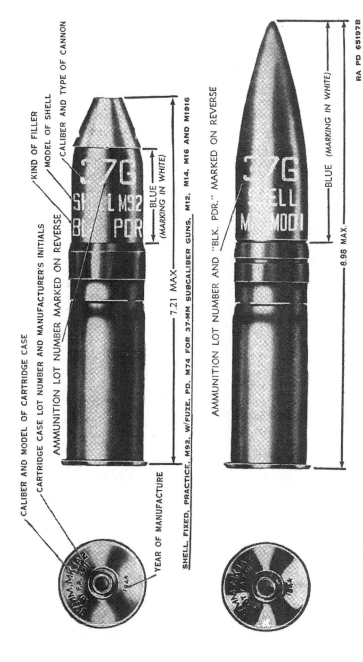

CALIBER AND MODEL OF CARTRIDGE CASE

CARTRIDGE CASE LOT NUMBER AND MANUFACTURER'S INITIALS

AMMUNITION LOT NUMBER MARKED ON REVERSE

KIND OF FILLER

MODEL OF SHELL

CALIBER AND TYPE OF CANNON

BLUE (MARKING IN WHITE)

7.21 MAX.

YEAR OF MANUFACTURE

SHELL, FIXED, PRACTICE, M92, W/FUZE, PD, M74 FOR 37-MM SUBCALIBER GUNS, M12, M14, M16 AND M1916

AMMUNITION LOT NUMBER AND "BLK. PDR." MARKED ON REVERSE

BLUE (MARKING IN WHITE)

8.98 MAX.

RA PD 65197B

SHELL, FIXED, PRACTICE, M63 - MOD 1, W/FUZE, BD, PRACTICE, M58 FOR 37-MM SUBCALIBER GUNS, M12, M13, M14, M16 AND M1916

Figure 102. Ammunition for 37-mm subcaliber guns.

148

Section XXIII. SIGHTING AND FIRE CONTROL EQUIPMENT

124. Arrangement of Sighting and Fire Control Equipment

a. APPLICATION. (1) The telescope mount M21A1 with panoramic telescope M12A2 and instrument light M19, the range quadrant M4A1 and the telescope mount M23 with instrument light M36 and elbow telescope M16A1D comprise the on-carriage sighting equipment used with the 105-mm howitzer M2A1 and 105-mm howitzer carriages M2A1 and M2A2. The same equipment with the exception of the telescope mount M23 and elbow telescope M16A1D are furnished for the howitzer mounts M4 and M4A1. The telescope mount M42 and elbow telescope M16A1C are furnished for the howitzer mounts M4 and M4A1.

(2) For a list of off-carriage sighting and fire control equipment see paragraph 6*e.* For description of the graphical firing table, refer to TM 9–524 or TM 9–526.

b. TELESCOPE MOUNT M21A1 WITH PANORAMIC TELESCOPE M12A2. The telescope mount M21A1 with panoramic telescope M12A2 are used for aiming the howitzer in azimuth for either direct or indirect fire.

c. RANGE QUADRANT M4A1. The range quadrant M4A1 is used to lay the howitzer in elevation.

d. TELESCOPE MOUNTS M23 AND M42 WITH ELBOW TELESCOPES M16A1D AND M16A1C. These mounts with their respective telescopes are used in laying the howitzer for elevation when using the two-man, two-sight system of direct fire on moving targets as described in paragraph 127*d.*

125. Telescope Mount M21A1

The telescope mount M21A1 (figs. 103 and 104) is bracketed to the left side of the howitzer cradle and is centered about a prolongation of the cradle trunnion passing through the centering hole in the bracket. The panoramic telescope and the telescope mount provide the sighting element for laying the howitzer in direction. The mount consists principally of longitudinal-leveling mechanism, a cross-leveling mechanism, an azimuth compensating mechanism, and a telescope socket (fig. 104). The azimuth compensating mechanism, including the actuating arm, bearing, and pivots, provides a means of transmitting the motion of the barrel to the mount so that any deviation in deflection, due to elevating the barrel, may be corrected. The cross- and longitudinal-leveling mechanisms provide a means of overcoming errors due to cant by leveling the mount to permit setting off true deflections on the panoramic telescope. A pair of indexes are provided on the rocker and actuating arm of the telescope mount (fig. 104) to indicate a coarse setting when the mount and the telescope are in the proper relation to the axis of the bore. The

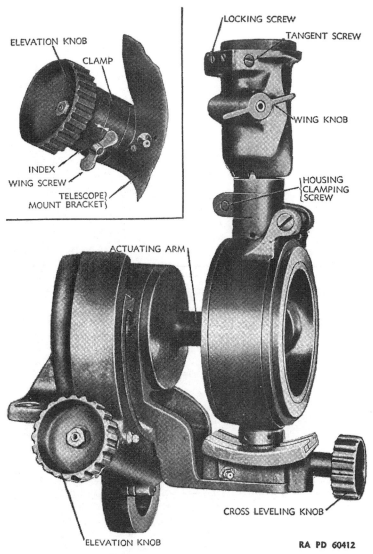

ELEVATION KNOB

CLAMP

INDEX

WING SCREW

TELESCOPE
MOUNT BRACKET

LOCKING SCREW

TANGENT SCREW

WING KNOB

HOUSING
CLAMPING
SCREW

ACTUATING ARM

CROSS LEVELING KNOB

ELEVATION KNOB

RA PD 60412

Figure 103. Telescope mount M21A1.

fine setting is obtained by matching indexes on the end of the elevation worm (fig. 103). The settings are secured against accidental movement by a clamp assembled to the end of the elevation worm and to the bracket. The panoramic telescope is secured in position in the socket by means of a retaining shaft which is rotated by a wing knob (fig. 103).

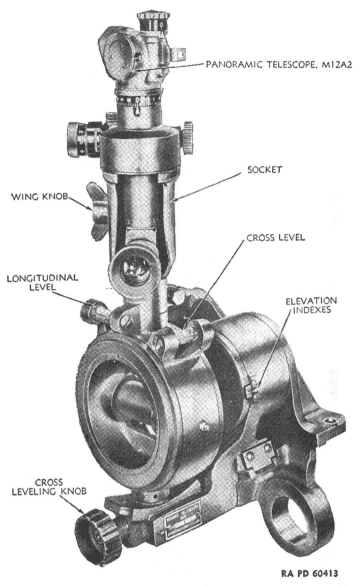

PANORAMIC TELESCOPE. M12A2

SOCKET

WING KNOB

CROSS LEVEL

LONGITUDINAL LEVEL

ELEVATION INDEXES

CROSS LEVELING KNOB

RA PD 60413

Figure 104. Telescope mount M21A1 with panoramic telescope M12A2.

151

126. Panoramic Telescope M12A2

a. The panoramic telescope M12A2 (fig. 105) is a 4-power fixed-focus vertical telescope of the periscopic type with a rotating head and azimuth mechanism by which the line of sight may be directed to any desired azimuth. It also has a movable prism which permits the line of sight to be elevated or depressed through a limited angle (±300 mils from the zero position) as required to keep the aiming point within the field of view. The line of sight is elevated or depressed by turning the elevation knob (fig. 105) located at the top of the panoramic telescope. The zero setting (at the center of the 600 mil movement) is indicated when the coarse elevation index lines on the rotating head and the index lines on the elevation micrometer are matched, but the angles of depression or elevation cannot be measured as no scales are provided. The azimuth scale consists of a circular ring graduated into 64 divisions of 100 mils each, and is numbered every 400 mils progressively from 0 to 3,200 in two consecutive semicircles. The motion of the azimuth worm knob, or the azimuth micrometer knob, is imparted to the rotating head and azimuth scale. The azimuth micrometer is clamped to the micrometer knob by the two screws in the end of the knob. Adjustment can be made by loosening these two screws and shifting the micrometer. The

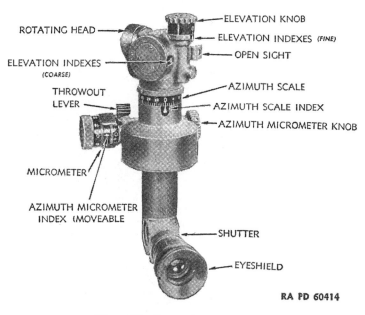

ROTATING HEAD
ELEVATION INDEXES (COARSE)
THROWOUT LEVER
MICROMETER
AZIMUTH MICROMETER INDEX (MOVEABLE

ELEVATION KNOB
ELEVATION INDEXES *(FINE)*
OPEN SIGHT
AZIMUTH SCALE
AZIMUTH SCALE INDEX
AZIMUTH MICROMETER KNOB
SHUTTER
EYESHIELD

RA PD 60414

Figure 105. Panoramic telescope M12A2.

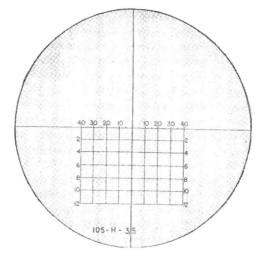

Figure 106. Reticle pattern, panoramic telescope M12A2.

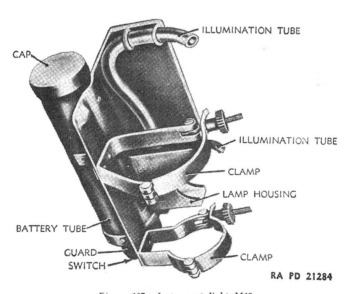

RA PD 21284

Figure 107. Instrument light M19.

micrometer is graduated into 100 divisions representing 1 mil each. The azimuth micrometer index (fig. 105) is adjustable in relation to a fixed scale graduated on the worm throwout lever. The latter scale is provided to set off deflections up to 20 mils right or left.

b. The reticle (fig. 106) is gridded with horizontal range lines indicating ranges from 0 to 1,200 yards in increments of 200 yards, and with vertical lines indicating deflections from 0 to 40 mils either right or left of the center (zero) line in increments of 10 mils. This type of reticle permits use of the telescope and its mount in the one-sight, one-man system of direct laying. The reticle is graduated for ranges corresponding to charge 5, HE shell M1, and is so marked. At short ranges (under 2,400 yards) the elevations for charge 7, HE shell M1, are almost exactly half those for charge 5. Hence, the reticle can be used for charge 7 by laying with half the true range.

c. The panoramic telescope M12A2 is constructed so as to utilize the 45-degree offset position of the telescope socket in the body of the mount. This permits the gunner to be in an easier position when looking through the instrument. The instrument light M19 (fig. 107) is provided for the illumination of the reticle, the scales, and the indexes of the panoramic telescope M12A2, and the elevation index and level bubbles of the telescope mount M21A1.

127. Operation

a. TELESCOPE EXTENSION. If the telescope extension is to be used, loosen the housing clamping screw (fig. 103) and lift out the telescope socket. Position the end of the extension in the housing and tighten the housing clamping screw. Loosen the clamping handle on the extension and position the telescope socket on the extension. Tighten the clamping handle.

b. TELESCOPE INSTALLATION. To place the telescope in its socket, remove the telescope from the carrying case, turn the wing knob on the mount to its extreme counterclockwise position, and place the telescope gently in the socket. Exert slight downward pressure to insure that the telescope is properly seated. Release the wing knob. Uncover both levels of the telescope mount. Operating procedure depends upon whether direct or indirect laying is to be employed, instructions for which are given separately.

c, DIRECT LAYING—ONE-MAN, ONE-SIGHT SYSTEM. Set the fine and coarse elevation indexes on the panoramic telescope rotating head at zero, and leave them in this position. Match the indexes on the actuating arm and rocker and on the end of the elevation worm of the telescope mount by means of the longitudinal-leveling knob, and clamp them in this position with the wing screw. Cross-level the telescope mount. Cross-leveling of the telescope mount is maintained throughout direct fire only as necessary to keep the range lines of the reticle approximately horizontal to prevent erroneous range applications. Set the azimuth micrometer scale to zero opposite the fixed index. Lateral deflections up to 40 mils either right or left may be read directly from the reticle, thus eliminating the necessity for using the azimuth mechanism.

d. DIRECT LAYING—TWO-MAN, TWO-SIGHT SYSTEM. The gunner follows the target with the appropriate lead, using the reticle of the panoramic telescope and the traversing handwheel on the howitzer carriage. The No. 1 cannoneer follows the target with the appropriate range line of the reticle in the elbow telescope M16A1D by rotating the elevating handwheel on the right of the howitzer carriage.

e. INDIRECT LAYING. (1) Set the deflection (firing angle) on the azimuth and azimuht micrometer scales. Disregard the red markings on the scale since the firing angle is always a clockwise angle.

(2) Center the bubbles in the cross- and longitudinal-level vials and keep them continuously centered during fire by means of the cross- and longitudinal-leveling mechanism knobs.

(3) Operate the traversing handwheel on the howitzer carriage so that the zero vertical line of the telescope reticle remains continuously on the aiming point. It is not necessary to bring the aiming point exactly on the zero horizontal line, but if the aiming point does not fall within the field of view, rotate the telescope's elevation knob until it does. This procedure is permissible in *indirect fire only.*

f. CHANGES IN DEFLECTION. With the deflection set on the telescope and the micrometer scale set to zero opposite the fixed index, rotate the azimuth micrometer index, by means of the azimuth worm knob, in the proper direction the required number of 1-mil divisions.

g. FOR TRAVEL. Turn the wing knob counterclockwise and lift out the telescope. Remove the telescope extension. Place the telescope and extension in the panoramic telescope case. Protect both levels by closing their covers.

128. Test and Adjustment

a. GENERAL. In general, the sighting equipment is correct in the following:

(1) *Direction.* If the deflection scales read zero when the line of sighting is in a plane parallel to the vertical plane passing through the axis of the bore.

(2) *Elevation.* If the algebraic sum of the site and elevation settings indicates the same angle above or below the horizontal that is measured with the gunner's quadrant on the leveling plates on top of the breech ring. In the case of direct laying with the elevation indexes of the panoramic telescope set at zero and with the elevation indexes of the sight mount in coincidence, when the planes of sighting through the zero range line of the panoramic and elbow telescopes is parallel to the axis of the bore.

(3) *Laying.* If there is no lost motion between the sighting and laying equipment and the tube. In direct laying, the range lines of the elbow and panoramic telescope must be kept horizontal in order that the proper range and angle of site will be laid on the target.

b. EQUIPMENT. Equipment used in testing the sights consists of bore sights and a gunner's quadrant. The target for bore sighting may be a terrain object approximately 2,000 yards away, or a test target at least 50 yards distant. Bore sighting may be done without the issue bore sights by removing the firing lock and sighting through the firing pin hole in the breechblock bushing, or through a brass cartridge case with the primer removed, using improvised cross hairs at the muzzle. The same testing target cannot be used for the various panoramic telescopes which may be encountered. Exercise care to use the target prepared for the telescope actually in use. Paragraph 133 provides the correct dimensions for testing target for the panoramic telescopes which may be used with this matériel.

c. LOST MOTION. (1) The worm gears of the sighting and laying equipment should be tested periodically for lost motion. The effect of small amounts of lost motion may be eliminated by habitually making the last movement always in the same direction. The last movement in setting and laying for deflection should be from left to right. The last movement in setting the scales should be in the direction of increasing the reading. Lost motion in the elevating mechanism of the carriage may be taken up by moving the elevating handwheel against the greatest resistance.

(2) If an appreciable degree of lost motion exists in the sighting and laying equipment, adjustment should be made without delay by ordnance personnel. The artillery mechanic is authorized to correct looseness of the panoramic telescope in the socket of the mount by an adjustment of the tangent screws. Exercise care that there is no undue pressure by either tangent screw. The panoramic telescope must seat firmly without binding.

d. LEVELING THE CARRIAGE. In bore sighting and testing, the trunnions should be leveled, and it is desirable that the piece be in the center of the traverse. The trunnions should be leveled by leveling the ground under the trails or by blocking the lower trail to the height of the higher trail. The leveling of the trunnions should be checked with a gunner's quadrant on the breech ring in conjunction with a machined steel plate or a piece of plate glass.

e. TELESCOPE MOUNT AND PANORAMIC TELESCOPE. Periodically, and whenever the mechanism is found to be out of adjustment, a detailed test and adjustment should be made as explained below.

(1) *Cross and longitudinal levels of the telescope mount.* Remove the telescope from the telescope mount and cross-level the telescope mount.

(*a*) Place a gunner's quadrant (set for zero elevation with the correction, if any, applied) on top of the telescope socket parallel to the axis of the bore, and center the quadrant bubble by means of the longitudinal-level knob. (A machined steel plate or a piece of plate glass placed on the locating surface of the mount will provide a surface for seating the gunner's quadrant.) Note the position of the bubble in the longitudinal level. If the bubble is not centered by over one graduation, adjustment may be made by responsible battery personnel or by the ordnance personnel.

(*b*) With the telescope mount vertical, place a gunner's quadrant (set for zero elevation with the correction, if any, applied) on the top of the telescope socket parallel to the axis of the howitzer trunnions, and center the quadrant bubble by means of the cross-level knob. If the cross-level bubble is not centered, adjustment may be made by responsible battery personnel or by the ordnance personnel, as in step (1)(*a*) above.

(2) *Elevation indexes of the telescope mount.* Level the howitzer with the gunner's quadrant and center the cross- and longitudinal-level bubbles. Clamp the elevation worm in position with the wing screw. The elevation indexes on the actuating arm and rocker and on the elevation worm of the mount should coincide. If they do not, adjust the actuating arm and rocker indexes by loosening the two screws on the adjustable index and slide the movable index into coincidence with the fixed index, tighten the screws, and recheck the level bubbles. To adjust the indexes on the elevation worm, loosen the clamping nut in the elevation knob and slip the index until it matches the index on the clamp. Tighten the nut in the elevating knob.

(3) *Panoramic telescope M12A2.* With the elevation indexes of the telescope mount in coincidence, the trunnions level, and the cross-level bubble centered, bore sight on the testing target or the terrain object (see par. 133).

(*a*) *For elevation.* With the elevation knob, place the zero range line (optical center) of the reticle on the corresponding line of the testing target or on the terrain object. The elevation micrometer and the elevation index should indicate zero. If the elevation micrometer does not indicate zero, loosen the screws in the end of the elevating knob and, holding the knob, slip the micrometer until the zero graduation coincides with the index. Then tighten the screws and recheck. If the coarse elevation index does not indicate zero, the adjustment should be made by ordnance personnel.

(*b*) *For direction.* With the azimuth worm knob, place the vertical line (optical center) of the reticle on the corresponding line of the testing target or on the terrain object. When the vertical line (optical center) of the reticle is placed on the corresponding line of the testing target or on the terrain object and zero is not indicated on either one or both of the scales, they may be brought to zero by loosening appropriate screws, slipping the scales into coincidence, and retightening.

129. Range Quadrant M4A1

a. DESCRIPTION. Range quadrant M4A1 (fig. 108) is mounted on the right side of the howitzer cradle and provides the means for laying the howitzer in elevation. The range quadrant is lighted by a built-in system supplied by four flashlight batteries, BA–30, which are mounted in a compartment of the quadrant. The range quadrant includes:

(1) An angle of site level for applying elevation and site.

(2) A cross level and leveling mechanism for applying elevation and site in a vertical plane.

(3) Angle of site and angle of elevation mechanisms which introduce and add algebraically their respective elements of data.

(4) Canvas cover M411 is provided for the range quadrant M4A1 and the telescope mount M23 with elbow telescope M16A1D. These covers should be used when it is necessary to have the instruments setup in position while not in actual use.

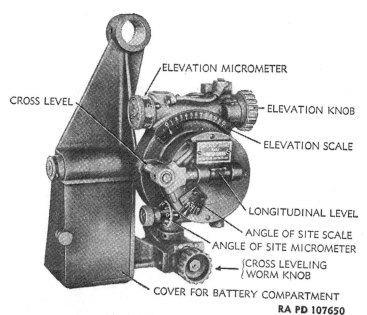

Figure 108. Range quadrant M4A1.

b. Operation. (1) To operate the range quadrant, either of the two following combinations of data may be employed:

(*a*) Angle of site in mils, angle of elevation in mils.

(*b*) Quadrant elevation in mils set as angle of elevation with the angle of site set at normal (300), since the quadrant elevation already includes any necessary angle of site.

(2) Having made the necessary settings in accordance with the combination selected from the above listing, cross-level the range quadrant by means of the cross-leveling knob. Then elevate or depress the howitzer until the longitudinal-level bubble is centered, checking the cross-level bubble at all times. The howitzer is then properly laid in elevation. Maintain both bubbles in their centered position as long as firing is continued.

(3) For travel, close the covers on the level vials.

c. Test and Adjustment. Level the howitzer and carriage horizontally (axis of bore and axis of trunnions). Center the cross-level bubble.

(1) *Elevation scales.* Center the longitudinal-level bubble, and note if the elevation scale and micrometer both read zero. If the elevation scale does not read zero, loosen the screws in the index plate and move the index opposite the zero graduation; tighten the screws and recheck. If the elevation micrometer does not read zero, loosen the three screws in the knob and, without moving the knob, slip the zero of the micrometer into coincidence with the index; tighten the screws and recheck. The scale and micrometer should read zero simultaneously.

(2) *Angle of site.* With the howitzer at zero elevation, the elevation scales at zero, the cross-level bubble centered, and the longitudinal-level bubble centered, the angle of site scale should read 3 (300) and the angle of site micrometer should read zero. If the angle of site scale does not read 3 (300), loosen the two nuts on the scale and move the figure 3 into coincidence with the index. If the angle of site micrometer does not read zero, loosen the nut in the micrometer knob and holding the micrometer knob so as to keep the bubble centered, move the zero into coincidence with the index. Tighten the nuts and recheck.

130. Telescope Mounts M23 and M42 with Elbow Telescopes M16A1D and M16A1C

a. Description of Mounts. The telescope mounts M23 (figs. 109 and 110) and M42 (fig. 111) are mounted on the upright extension of the range quadrant, and position the elbow telescope between the range quadrant and the howitzer. The telescope mount M23 which is furnished with the howitzer carriages M2A1 and M2A2 is being equipped with a clamp for the instrument light M36 (fig. 112). This mount contains mechanism for adjustment in elevation and for leveling the reticle lines.

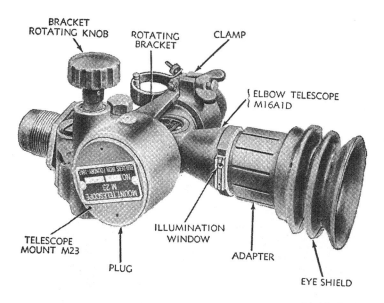

BRACKET
ROTATING KNOB

ROTATING
BRACKET

CLAMP

{ ELBOW TELESCOPE
{ M16A1D

TELESCOPE
MOUNT M23

ILLUMINATION
WINDOW

PLUG

ADAPTER

EYE SHIELD

RA PD 60419

Figure 109. Telescope mount M23 with elbow telescope M16A1D.

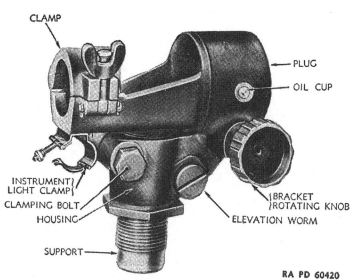

CLAMP

PLUG

OIL CUP

INSTRUMENT}
LIGHT CLAMP{

CLAMPING BOLT

HOUSING

SUPPORT

{BRACKET
{ROTATING KNOB

ELEVATION WORM

RA PD 60420

Figure 110. Telescope mount M23—top view.

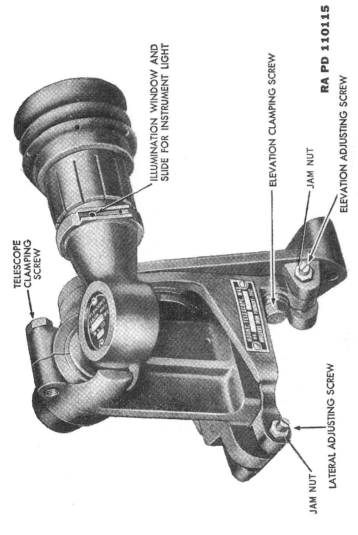

TELESCOPE
CLAMPING
SCREW

ILLUMINATION WINDOW AND
SLIDE FOR INSTRUMENT LIGHT

ELEVATION CLAMPING SCREW

JAM NUT

ELEVATION ADJUSTING SCREW

RA PD 110115

LATERAL ADJUSTING SCREW

JAM NUT

Figure 111. Telescope mount M42, with elbow telescope M16A1C.

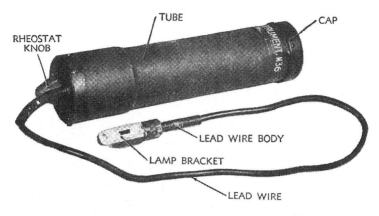

TUBE

CAP

RHEOSTAT
KNOB

LEAD WIRE BODY

LAMP BRACKET

LEAD WIRE

RA PD 60421

Figure 112. Instrument light M36.

The telescope mount M42 is provided with tangent screws for adjusting it in elevation and azimuth; no provision is made for leveling the reticle lines.

b. DESCRIPTION OF ELBOW TELESCOPE. (1) The elbow telescopes M16A1D and M16A1C (fig. 113) are 3-power, fixed-focus instruments. The telescopes are identical with the exception of their reticle patterns.

(2) The reticle pattern for elbow telescope M16A1D (fig. 114) has range lines only representing elevations for ranges from 0 to 2,200 yards in steps of 200 yards, with the 0-range (N) line passing through the optical axis. A dot in the center of the "N" range line is used for bore sighting. The reticle graduations are based upon data from Firing Table 105–H–3, changes No. 8, fired with a muzzle velocity of 1,250 feet per second, minus 0.4-mil jump. The inscription "105 MM HOW. H.E.A.T., M67" at the upper edge of the reticle indicates the ammuntion, and the number "7673922" at the lower edge is the part number of the reticle.

(3) The reticle pattern for the elbow telescope M16A1C as seen through the telescope is shown in figure 114. It is designed for use with two types of ammunition. The right-hand portion of the reticle pattern is based upon data from Firing Table 105–H–3, Part 2G for the H.E. shell, M1, charge 7, fired with a muzzle velocity of 1,550 feet per second, plus 0.4-mils jump. The left-hand portion of the reticle pattern is based upon data from Firing Table 105–H–3, C 8, for H.E.A.T., shell M67, fired with a muzzle velocity of 1,250 feet per second, plus 0.7 mils jump. The cross at the top of the pattern represents zero range and zero deflection for bore sighting. The inscription "105 MM HOW." at the upper edge of the reticle indicates the howitzer with which the reticle is used, and the inscription "H.E.A.T." and "VII" just above

the pattern designates the portion of the reticle to be used with each type of ammunition. The number "7673921" at the lower edge is the part number of the reticle. In all other respects the reticle is similar to the pattern for the M16A1D model.

c. OPERATION. (1) To place the elbow telescopes and telescope mount M23 in operation, clamp the telescope in the bracket of the telescope mount, having first removed all dust and dirt from the bearing surfaces. Insure that the telescope is fully inserted, that the projecting lug fits the mating opening on the telescope mount, and that the wing nut is securely tightened. Level the reticle lines with respect to the field by means of the bracket rotating knob on the upper portion of the telescope mount. This setting must be done by observation, as there is no mechanical means provided for determining the setting. The elbow telescope M16A1C is secured in its mount at all times. Attach the instrument light lamp bracket.

(2) *Direct laying.* The howitzer is elevated or depressed until the base of the target appears on the reticle graduation corresponding to the target range. Turn the illumination on if necessary, with the rheostat knob. The howitzer is then laid for range and angle of site, automatically.

(3) *For travel.* The elbow telescope used with the telescope mount M23 should be removed from its mount and placed in its proper place

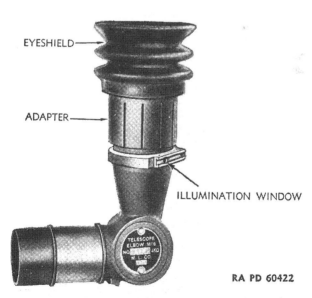

RA PD 60422

Figure 113. Elbow telescope M16A1D or M16A1C—top view.

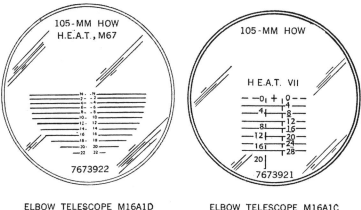

ELBOW TELESCOPE M16A1D   ELBOW TELESCOPE M16A1C

RA PD 110116

Figure 114. Reticle patterns, elbow telescopes M16A1D and M16A1C.

in the carrying case attached to the shield. The elbow telescope used with the telescope mount M42 is left in place on the mount.

d. Test and Adjustment. With the axis of the trunnions leveled, bore sight on the test target or the distant terrain object. By observation, level the reticle with respect to the field by means of the bracket rotating knob (M16A1D model only). If the N cross-line of the M16A1D model or the cross of the M16A1C model is not on its line of the test target, put it on in the following manner:

(1) *Telescope mount M23.* Loosen the clamping bolt (fig. 110), and with a screw driver turn the elevation worm to bring the "N" range line of the reticle in coincidence with the proper line of the testing target or the terrain target. Tighten the clamping bolt, and recheck. There is no adjustment provided for deflection.

(2) *Telescope mount M42.* Loosen the elevation clamping screw and the two jam nuts on each elevation adjusting screw and on each lateral adjusting screw (fig. 111). With the aid of a screw driver turn the adjusting screw until the cross of the reticle falls on the aiming point of the testing target or on the terrain target. See that opposing screws are tight and then tighten the jam nuts and the elevation clamping screw. Check the adjustment.

e. Care and Preservation. (1) Keep the instrument protected, when not in use, with the canvas cover M411 provided to prevent moisture, dirt, and grit from accumulating on the locating surfaces.

(2) Refer to paragraph 134 for general instructions pertaining to care and preservation of instruments. No attempt should be made to rotate the elevation worm without first loosening the clamping bolt.

131. Fuze Setter M22 or M14

a. DESCRIPTION. (1) *Fuze setter M22.* (*a*) The fuze setter M22 (fig. 115) is a hand operated dialed instrument for setting the mechanical time fuzes. A time scale and corrector scale with corresponding indexes record the fuze number corresponding to the firing table figure for the desired time of flight of the projectile. (Fuze numbers or fuze seconds are angular measurements and are not directly proportioned to the time of flight.) The setting is locked with two wing screws so that any

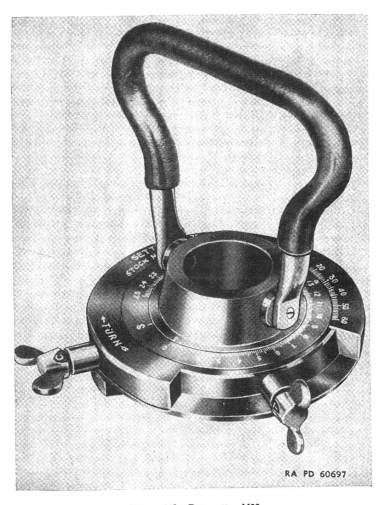

RA PD 60697

Figure 115. Fuze setter M22.

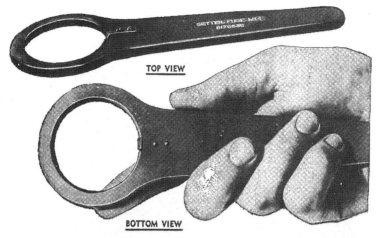

TOP VIEW

BOTTOM VIEW

RA PD 102669

Figure 116. Fuze setter M14.

number of fuzes may be set, even in darkness, until a new setting is required. The fuze setter is contoured to fit over the point of the fuze after the fuze safety pin has been removed.

(*b*) The time scale is graduated in 0.1-second intervals from 0 to 25, and indicates the desired fuze setting plus or minus the corrector setting. The letter "S" on the scale indicates the SAFE setting for the fuze when the corrector scale is set at 30 (normal).

(*c*) The graduations on the corrector scale are called points. The scale is graduated in 1-point (0.1-second) intervals from 0 to 60 points with 30 as normal. The graduations represent corrections in time from 0 to 3 seconds either side of normal for increasing or decreasing the time of burst from the time indicated on the time scale.

(*d*) To insure accuracy in setting scales, look squarely at the graduations and indexes.

(*e*) The wing screws which lock the time scale and corrector scale settings are marked "T" and "C" respectively (fig. 115). Tighten the "C" screw and loosen the "T" screw when setting time values. Then, turn the handle until the time scale index alines with the desired graduation; tighten the "T" screw. To set the corrector values, loosen the "C" screw and turn the corrector scale until its index alines with the desired graduation; tighten the "C" screw.

(*f*) The handle which was used to set the time values is also used to turn the fuze setter when setting fuzes.

(*g*) The carrying case M66 is provided for the fuze setter M22.

(2) *Fuze setter M14.* The fuze setter M14 (fig. 116) is designed as

a flat handled wrench having a circular tapered hole at one end to fit the fuze and a key which protrudes through the hole to engage the slot of the fuze. After the fuze safety pin has been removed the fuze setter is positioned on the fuze and turned clockwise (increasing direction) until the index mark on the fuze alines with the required time setting on the fuze scale.

b. To Set a Fuze. (1) Place the fuze setter, with the scales set to the desired values, over the point of the fuze. Press down firmly on the fuze setter and at the same time rotate the fuze setter clockwise until the lug engages the fuze ring. Continue rotation until the pawl seats in the fuze body. The setting is complete when further rotation becomes impossible.

(2) If it is desired to reset a fuze to the SAFE position, set the time scale to "S" and the corrector scale to the "30" and proceed as above.

Caution. Before setting a fuze, make sure that the "T" and "C" screws are tight to prevent any slipping of the scale indexes when the handle of the fuze setter is rotated. When setting fuzes always rotate the fuze setter in a clockwise direction. When removing the fuze setter from the fuze, lift it straight off without rotating it to prevent disturbing the setting of the fuze.

c. Tests and Adjustments. Test the fuze setter M22 on a dummy, inert, or live fuze for correct setting of the fuze, and for smooth operation. Set the corrector scale to normal (30), and set in some value on the time scale. Cut the fuze. The time setting of the fuze should agree with the setting on the fuze setter time scale. If the settings do not agree, repeat the operation with a different time value to make sure there was no slippage. Note engagement with fuze and any tendency to stick or bind. If the fuze setter fails to operate properly, turn it in for repairs by ordnance maintenance personnel. No adjustments by the using personnel are permitted.

Caution. Where a live fuze is used, the precautions normally observed in handling ammunition must be followed. Remove the safety wire or cotter pin carefully for the test. After the test has been made, return the fuze setting to the "S" or "SAFE" setting, and replace the safety wire or cotter pin. When checking the accuracy of the fuze setter by cutting trial fuzes, no fuze should be cut more than twice. The fuze from a dud must never be used. Further precautions are described in TM 9-1900.

132. Bore Sight (fig. 39)

a. General. The bore sight is used to indicate the direction of the axis of the bore of the piece, for orientation purposes. Each bore sight is composed of a breech element and a muzzle element.

b. DESCRIPTION. The breech bore sight is a metal disk which fits accurately in the powder chamber of the howitzer. The model of the howitzer for which it is to be used is engraved on the disk. The muzzle bore sight includes a quantity of black linen cord, to be stretched tightly across the muzzle, vertically and horizontally, in the score marks thereon, and a web belt to be buckled around the muzzle to hold the cord in place.

c. OPERATION. With the two elements in place, look through the aperture in the breech bore sight; the direction of the axis is indicated by the cord intersection.

133. Testing Targets (figs. 117 and 118)

The testing target is used during the bore-sighting operation for the alinement of sights with the axis of the bore. The aiming points are plainly designated. It is *essential* that the proper aiming points be selected for the matériel and equipment employed, and that the target be positioned vertically in a transverse vertical plane when in use. The normal distance from the muzzle at which the target should be located is about 50 yards. The targets for use with the different panoramic telescopes are not interchangeable and care should be taken that the correct target is used. The testing target for use with the M2A2 carriage is shown in figure 117 and the one for use with the mount M4 or M4A1 is shown in figure 118.

134. Care and Preservation

a. GENERAL. (1) The instructions given hereunder supplement instructions pertaining to individual instruments included in preceding paragraphs.

(2) Fire control and sighting instruments are, in general, rugged and suited for the purpose for which they have been designed. They will not, however, stand unnecessarily rough handling or abuse; inaccuracy or malfunctioning will result from such mistreatment.

(3) Disassembly and assembly by the using arm are permitted only to the extent described in the paragraphs dealing with the individual instruments. Where stops are provided to limit rotation, no attempt should be made to force the mechanism beyond these limits.

(4) Canvas cover M412 has been designed to provide protection from weather damage for the panoramic telescope M12A2 and telescope mount M21A1, when it is necessary to have the instruments set up in position while not in actual use. If it is not necessary to have the instruments set up, keep them in the carrying cases provided or in the condition indicated for traveling. Keep the panoramic telescope and elbow telescope M16A1D in the panoramic telescope case. A bracket is

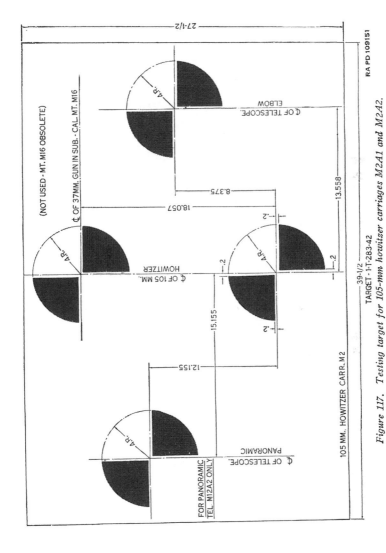

Figure 117. Testing target for 105-mm howitzer carriages M2A1 and M2A2.

169

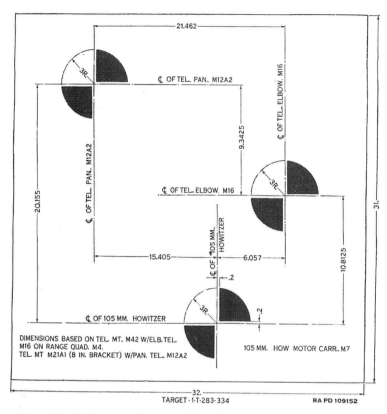

Figure 118. Testing target for 105-mm howitzer motor carriages M7 and M7B1.

provided in the panoramic telescope case for carrying the elbow telescope.

(5) Any instruments which indicate incorrectly or fail to function properly after the authorized tests and adjustments have been made are to be turned in for repair by ordnance personnel. Adjustments other than those expressly authorized in the paragraphs dealing with the individual instruments are not to be performed by the using arm.

(6) Many worm drives have worm throw-out levers to permit rapid motion through large angles. When using these mechanisms, it is essential that the throw-out lever be fully depressed to prevent injury to the worm and gear teeth.

(7) Do not attempt to turn the elevation knob on the mount while the wing screw is tightened.

b. Optical Parts. (1) To obtain satisfactory vision, it is necessary that the exposed surfaces of the lenses and other parts be kept clean and dry. Etching of the surface of the glass can be prevented or greatly retarded by keeping the glass clean and dry.

(2) *Under no condition* will polishing liquids, pastes, or abrasives be used for polishing lenses and windows.

(3) For wiping optical parts, use only lens tissue paper. The use of cleaning cloths is not permitted. To remove dust, brush the glass lightly with a clean camel's-hair brush. Rap the brush against a hard body in order to knock out the small particles of dust that cling to the hair. Repeat this operation until all dust is removed. With some instruments, an additional brush with coarse bristles is provided for cleaning mechanical parts. It is essential that each brush be used only for the purpose intended.

(4) Exercise particular care to keep optical parts free from oil and grease. Do not wipe the lenses or windows with the fingers. To remove oil or grease from optical surfaces, apply liquid lens cleaning soap with a tuft of lens tissue paper, and rub gently with clean lens paper. Wipe dry with clean lens tissue paper. If liquid lens cleaning soap is not available, breathe heavily on the glass (provided the temperature of the surrounding air is above 32° F.) and wipe off with clean lens paper. Repeat this operation several times until clean. Under no circumstance will a hose, either normal-pressure or high-pressure, be used in cleaning any sighting or fire control instruments.

(5) Moisture due to condensation may collect on the optical parts of the instrument when the temperature of the parts is lower than that of the surrounding air. This moisture, if not excessive, can be removed by placing the instrument in a warm place. No direct heat from strongly concentrated sources should be applied, as it may cause unequal expansion of parts, thereby resulting in breakage of optical parts or inaccuracies in observation.

(6) Many optical instruments are composed of elements which have been coated, that is, upon which a reflection reducing film of magnesium fluoride has been deposited. The present method of coating provides a film which is hard and durable, and will not be destroyed by the cleaning method described above. Optical coatings can be identified by the color of the reflected light. The reflection from a coated surface is much reduced in intensity and has a distinctive purplish-red or bluish tinge. If the coating shows indications of having rubbed off in certain areas, it is probably a soft coating and extreme care must be taken in cleaning to do as little rubbing as possible. A coating which is partially rubbed off should not be removed, since the areas which remain coated are still effective in reducing reflections.

c. Lubrication. (1) Lubricating fittings on the telescope mount

M21A1 and range quadrant M4A1 are not to be used as a means of lubricating the matériel, since the internal mechanisms are lubricated at assembly.

(2) Monthly, apply 1 to 2 drops of lubricating oil for aircraft instruments and machine guns, to bracket rotating worm and bracket trunnion on the telescope mount M23 through oiler located on top of housing.

(3) Monthly, saturate with preservative lubricating oil (special) the felt washers on the range quadrant and telescope mount. The standard oiler can be used for this purpose.

(4) Under normal conditions, wipe exposed bearing surfaces, such as the lugs and collar of the panoramic telescope and locating surfaces of the telescope mount M21A1, clean and renew oil film daily. Reduce this interval, if necessary, under extremely high humidity, salt air, or moisture conditions.

Caution. Apply oil sparingly when operating in sandy or dusty areas, and during extremely low temperatures.

Use preservative lubricating oil (special) for all temperatures, except for high humidity, salt air, or moisture conditions above $+32°$ F. Use preservative lubricating oil (medium) for extremely high humidity, salt air, or moisture conditions above $+32°$ F. Keep the exterior surfaces of the telescope mounts and range quadrant free of dirt and excess lubricants.

135. Cold Weather Operation

a. GENERAL. Sighting and fire control equipment should not be exposed to sudden changes in temperature because of the dangers of condensation, and the effects of sudden lowering of temperatures on the accuracy of the equipment. If a piece of equipment is going to be used outside at low temperature, it should stay outside and not be stored at room temperature and brought outside when it is to be operated.

b. CONDENSATION ON OCULARS. (1) When using optical instruments in cold weather, the operator should be careful not to breathe on the eyepieces. When the breath hits the lenses, the moisture in the warm breath condenses on the lenses and fogs them. This moisture will then freeze, making it impossible to observe with the instrument.

(2) There is no antifog solution that is satisfactory for use on lenses of optical instruments at low temperatures. Some solutions prevent fogging, but they striate on the lens making observation difficult or impossible.

(3) A face mask is the most satisfactory method of keeping the breath away from the eyepieces. The face mask can be of any type, as long as it directs the breath away from the lens or absorbs the moisture from the breath. A serviceable face mask can be made from a piece of chamois or gauze tied across the face just below the eyes. This will not

only protect the operator's face from the wind but will direct the breath downward away from the lens. This type mask must be changed periodically to avoid freezing the face.

(4) Optical surfaces should be cleaned in cold weather by using a lens tissue paper moistened with a few drops of ethyl alcohol. If alcohol is not available, a dry lens tissue is satisfactory. Alcohol should never be applied directly to the surfaces, as it may injure the sealing compound.

c. PROTECTION OF TELESCOPES. (1) Snow will collect in uncovered eyepieces and objective tube sunshades rendering the instrument useless until the snow is removed. Do not try to blow the snow out of these parts or wipe it out with gloves or the bare hands. This will melt some of the particles of snow and they will freeze on the lenses, causing further difficulty. Use a small, stiff brush or small. rubber bulb and nozzle to remove the snow.

(2) The panoramic and elbow telescope should be covered while not in use. Place a cloth cover over the entire instrument. A cloth cover is better than an airtight cover, such as a leather cover because the cloth cover allows breathing of the air in contact with the lens and thereby prevents condensation on the lens if the instrument is subjected to a lower temperature. The covers can be made with a spring, elastic, or drawstring at the mount so they can be held in place and also easily and quickly removed.

(3) A temporary method of keeping snow out of eyepieces and objective tube extensions is putting loose wads of lens tissue paper into them when the instrument is not in use. These wads can be removed easily.

Section XXIV. SUBCALIBER EQUIPMENT

136. Purpose

Subcaliber equipment, which is used for training purposes only and is not taken into the theater of operations, consists of the 37-mm subcaliber gun M13 and equipment. It is used to provide more extensive training in laying and firing the 105-mm howitzer matériel than would be permissible with the standard 105-mm ammunition. The use of smaller bore ammunition prevents wear on the regular piece during practice and is less costly. Although the actual handling, loading, and range obtained are different, the results obtained in elevating, traversing, sighting, and similar operations are adequate for instructional purposes. The use of 105-mm drill ammunition provides training in handling and loading standard size ammunition.

137. Description and Functioning

a. GENERAL. The 37-mm subcaliber gun M13 is an interior type gun which is positioned in the bore of the 105-mm howitzer (fig. 119). The subcaliber gun consists of a tube, blast deflector, extractor, lock assembly, and cap screw. A collar near the breech end of the tube and the last deflector screwed onto the muzzle end of the tube support and aline the tube in the chamber and bore of the howitzer. The tube is secured in the howitzer by a welded lock assembly which fits in a vertical keyway at the rear of the subcaliber tube and fastens to the lower left side of the breech ring by means of a cap screw passing through a clamp on the lock assembly. The extractor, a sliding cylindrical sleeve which encases the rear end of the tube, completes the subcaliber gun. The rear end of the extractor is partly cut away and is counterbored for the rim of a 37-mm cartridge.

b. SUBCALIBER TUBE. The tube is 29.11 inches, or 20 calibers long. It has a chamber for a 37-mm cartridge and the bore is rifled with a uniform right-hand twist (one turn in 25 calibers).

c. EXTRACTOR. A rim, extending more than halfway around the rear of the extractor, is engaged by the extractor of the howitzer in the same manner as the rim of a 105-mm cartridge case. The left rear portion of the extractor is cut away to clear the lock assembly during extraction. Two slots in the right cylindrical wall reduce the weight.

d. PERCUSSION. Percussion is obtained by operation of the firing mechanism of the 105-mm howitzer.

138. Installation of Subcaliber Gun

Assemble the blast deflector to the muzzle end of the gun if not already assembled. Open the breech of the howitzer. Place the subcaliber extractor over the rear end of the subcaliber tube, alining the keyway for the lock assembly with the open side of the extractor. Insert the subcaliber gun into the bore of the howitzer so that the vertical keyway is toward the left and slightly to the rear of the forward face of the breech recess. Insert the lock assembly into the left side of the breech recess with the key of the lock fitting into the keyway of the subcaliber tube, and the lower leg of the clamp under the bottom wall of the breech ring. Push the lock and subcaliber tube forward until the lock rests in position against the left forward face of the breech recess. Tighten the cap screw to retain the lock and subcaliber tube. Note that as the sub-caliber tube is pushed into position the subcaliber extractor is held to the rear by the howitzer extractor. Operate the breech mechanism to test the action of the subcaliber extractor.

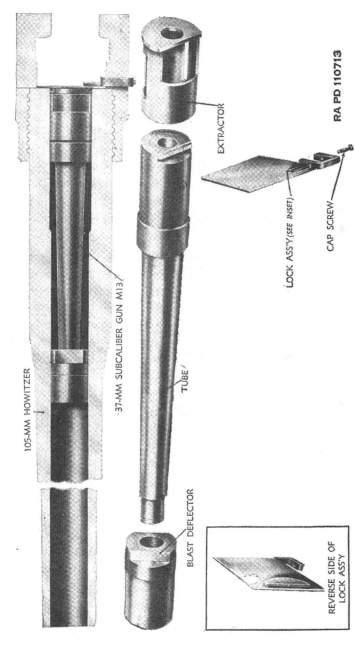

105-MM HOWITZER

37-MM SUBCALIBER GUN M13

TUBE

BLAST DEFLECTOR

EXTRACTOR

LOCK ASS'Y (SEE INSET)

CAP SCREW

REVERSE SIDE OF LOCK ASS'Y

RA PD 110713

Figure 119. 37-mm subcaliber gun M13.

175

139. Removal of Subcaliber Gun

Open the breech. Loosen the cap screw that secures the lock to the breech ring, pull the entire subcaliber gun to the rear by means of the extractor, and remove the subcaliber gun from the bore of the howitzer. The blast deflector should remain assembled to the subcaliber tube to protect the threads.

140. Operation of Subcaliber Gun

a. LOADING. (1) Open the breech of the howitzer. This will move the extractor of the subcaliber gun to the rear by the engagement of its rim with the lip of the 105-mm howitzer extractor.

(2) Insert a 37-mm cartridge into the counterbored hole of the subcaliber extractor, and position it in the chamber of the subcaliber tube (fig. 120). Close the breech of the howitzer.

b. FIRING. (1) Operate the firing mechanism of the 105-mm howitzer (par. 19).

(2) In case of failure to fire, proceed as follows:

(*a*) Check for proper operation of howitzer (par. 56*a*).

Figure 120. 37-mm subcaliber gun M13 (loaded) mounted in 105-mm howitzer M2A1—breech open.

(*b*) Attempt to fire twice more. If it still fails to fire, wait 30 seconds and open the breech. The safest time to remove the round from a hot chamber is between 30 and 45 seconds after the initial misfire occurred.

(*c*) Examine the primer. If the indent is normal, immediately segregate the round for the inspection of the ordnance officer in charge and the preparation of the reports required by AR 750-10. If there is no indent or the indent is light, proceed as in paragraph 56*c*.

c. EXTRACTING. (1) Open the breech of the 105-mm howitzer. This will pull the subcaliber extractor to the rear sharply by the action of the extractor of the 105-mm howitzer. At the end of the rearward movement, the subcaliber extractor is stopped by the forward face of the lock assembly, but the 37-mm cartridge case continues to move rearward by its own momentum. Complete ejection may not be accomplished but the cartridge case is moved back sufficiently to be removed by hand.

(2) In case of failure to extract, see paragraph 57. If the malfunction is not corrected by the procedure outlined therein, check the subcaliber extractor (fig. 119) for damage or excessive wear. If malfunction is due to the subcaliber gun, notify ordnance maintenance personnel.

d. UNLOADING. (1) A round, once loaded, will be fired rather than unloaded, unless military necessity or safety precautions dictate otherwise. The complete round is unloaded by opening the breech, catching the round, and segregating it for the inspection of the local ordnance officer to determine whether or not it can be reused.

(2) If the projectile becomes separated from the cartridge case when the breech is opened, notify ordnance maintenance personnel for its removal.

141. Preventive Maintenance of Subcaliber Gun

a. The procedure for cleaning, lubricating, and maintaining the subcaliber gun bore and chamber under either usual or unusual climatic conditions are the same as for the 105-mm howitzer bore and chamber as outlined in section XI at intervals as prescribed in paragraph 53. The procedure for servicing the subcaliber extractor and lock are the same as for the howitzer breech mechanism.

b. Oil the bore of the 105-mm howitzer lightly before installing the subcaliber gun. Before firing, wipe dry the bore of the subcaliber gun, but do not wipe dry the bore of the howitzer.

c. Note general appearance and smoothness of operation of extractor. Inspect parts for wear, burs, corrosion, or other damage, and correct any deficiencies.

d. Erosion of the tube is excessive and the accuracy life of the cannon is decreased by fast rate of firing for prolonged periods without proper cooling. See paragraph 6*d* for data pertaining to subcaliber equipment.

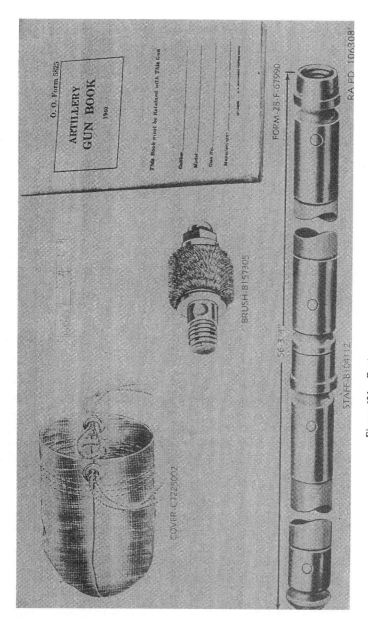

Figure 121. Equipment for gun, subcaliber, 37-mm, M13.

178

142. Organizational Spare Parts, Tools, and Equipment

a. SPARE PARTS. No spare parts are authorized for the 37-mm sub-caliber gun M13.

b. TOOLS AND EQUIPMENT. A set of organizational tools and equipment is supplied to the using arm for maintaining and using the matériel. This set contains items required for disassembly, assembly, cleaning and preserving the 37-mm gun matériel. Tools and equipment should not be used for purposes other than prescribed and, when not in use, should be properly stored in the chest provided for them.

c. LIST OF TOOLS AND EQUIPMENT. Tools and equipment for the 37-mm gun matériel are listed in WD Catalog ORD 7 SNL C–33, Section 16, which is the authority for requisitioning replacements. Table IV lists specially designed tools and equipment for use with the 37-mm subcaliber gun M13, and is for information only.

Table IV

Item	Identifying No.	References Figure	References Paragraph	Use*
Brush, bore, 37-mm, M8...	B157305	121	140	
Cover, bore, brush, 37-mm, M530	C7225002	121		
Cover, gun book, M539	C7228906		2	
Form, government, War Dept., artillery gun book, O.O. No. 5825 (blank)...	28–F–67990	121	2*a*	
Staff, cleaning	B104112	121	140	Used with bore brush.

*Where the use of an item is not indicated the nomenclature is self-explanatory.

APPENDIX I

STORAGE AND SHIPMENT

1. Preparation of Matériel for Shipment

a. UNCRATED MATÉRIEL. (1) Processing of the 105-mm howitzer M2A1 and carriages M2A1 and M2A2 for zone of interior shipments is *not required*, other than that the consignor of a shipment is responsible for the following (refer to SB 9–4):

(*a*) Providing to the shipper, matériel in a serviceable condition, properly painted and lubricated.

(*b*) All artillery gun bores, breech mechanisms, and exterior bearing surfaces must be coated with a thin-film of rust-preventive compound (light).

(2) If, during the course of shipment, operations will embrace deep water fording, matériel must be prepared in accordance with TM 9–2853.

b. CRATED MATÉRIEL. For crating and loading matériel for airplane or truck movement, refer to paragraphs 5, 6, and 7, this appendix.

2. Preparation of Matériel for Storage

a. GENERAL. (1) Matériel received for storage already processed for domestic shipment as indicated on "Vehicle Processing Record Tag" will not be reprocessed unless the inspection preparatory to or during storage reveals corrosion, deterioration, etc.

(2) Completely process matériel before storage (*b* through *l* below), if it is determined from the previous storage processing recorded on the tag that such has been rendered ineffective by operation, use or damage to the matériel, or upon receipt of matériel directly from manufacturing facilities.

b. MATERIALS REQUIRED. The required materials listed in Table V

Table V

ACID, phosphoric, metal conditioner, concentrated, wipe-off type, type II	OIL, lubricating, preservative (medium)
BARRIER, waterproof (Type C–1 and E–2)	PAPER, flint (sandpaper) grade 0 to grade 3.
COMPOUND, rust preventive (light and heavy)	WRAPPING, greaseproof (Type I, grade C)
COMPOUND, sealing, tape	

for preparation of matériel for storage are in addition to those listed in section XI.

c. RECEIVING INSPECTIONS. (1) When matériel is out of use, it must be turned over to ordnance personnel or placed in a storage status for periods not to exceed 90 days. Storage of matériel for periods in excess of 90 days will normally be handled by *ordnance personnel only.*

(2) Immediately upon receipt of matériel for storage, it must be inspected and serviced as prescribed in section IV. Make a systematic inspection and replace or repair all missing or broken parts. If repairs cannot be made prior to placing matériel in storage, attach a tag to the matériel specifying the repairs needed and make a written report of these items to the *officer-in-charge* of the matériel.

d. CLEANING. Prior to the application of preservatives and protective wrapping, thoroughly clean the howitzer and carriage as described below. Give special attention to bearing surfaces, revolving parts, springs, screw threads, gear teeth, and exterior surfaces, as well as the interior of the breech ring and the bore of the howitzer.

(1) *Cleaning nonrusted surfaces.* Follow cleaning instructions outlined in section XI.

(2) *Cleaning rusted surfaces.* Clean all metal surfaces that have become rusted or pitted as follows:

(*a*) *Unpainted metal surfaces.*
   1. Use crocus cloth for removing rust from finished surfaces.
   2. Use aluminum-oxide abrasive cloth for removing rust from unfinished surfaces where slight removal of metal will not affect the functioning of the part.
   3. Use type II, wipe-off type, concentrated, metal conditioner, phosphoric acid for removal of rust from unfinished surfaces where pits are too deep to be removed with aluminum-oxide abrasive cloth.

(*b*) *Painted metal surfaces.*
   1. Remove paint from rusted area using flint paper (sandpaper) grade 0 to grade 3.
   2. Remove rust as prescribed in (*a*) above and repaint.

e. LUBRICATION. The matériel will be completely lubricated before rail shipment or storage in accordance with current lubrication order. Refer to FM 21–6 for listing of current lubrication orders.

f. APPLICATION OF PRESERVATIVES AND PROTECTIVE WRAPPINGS. Apply preservatives immediately after cleaning and drying, as a rust stain will form if matériel is handled between operations. Apply rust preventive compound (light) and (heavy) hot, in order to obtain sufficient fluidity to adhere to the metal surfaces. This is best accomplished by placing the compound container in a vessel of water and heating.

Note. The maximum temperature to which rust preventive compound (light) may be heated is 150° F., and rust preventive compound (medium) may be heated to not more than 180° F.

Application of a flame directly to the compound container must be avoided, as overheating will destroy the protective qualities of the compound and may create a fire hazard. For description of preservatives used herein and method of application, refer to TM 9–850.

(1) *Howitzer tube.* (a) Swab the howitzer bore thoroughly, using a ramrod and clean cloths soaked in heated rust-preventive compound (light).

(b) Cover the muzzle with type I, grade C greaseproof wrapping material and secure with nonhygroscopic adhesive tape (fig. 122). Place a corrugated paper disk over the muzzle end and position a plywood disk against the corrugated paper disk. Secure with nonhygroscopic adhesive tape. Inclose the entire muzzle end with type E–2 waterproof barrier material, and thoroughly secure to the howitzer with nonhygroscopic adhesive tape. Apply tape sealing compound over the tape to protect it against the elements.

(2) *Breech mechanism.* (a) Apply heated rust-preventive compound (light) to the interior parts of the breech mechanism, and assemble to the howitzer in accordance with assembly procedure outlined in section XV. Before closing the breech, make a thorough inspection to insure that all unpainted metal surfaces are coated with rust-preventive compound (light). Seal the assembled breech mechanism in the breech ring with rust-preventive compound (heavy).

(b) Thoroughly inspect the exterior of the breech to insure that all exposed unpainted metal surfaces have been coated with rust-preventive compound (heavy). Seal the breech with two layers of type I, grade C greaseproof wrapping material. Secure with nonhygroscopic adhesive tape and apply tape sealing compound over tape.

(3) *Bolts and nuts.* Apply a coating of rust-preventive compound (light) to exposed threaded ends of bolts and threads of nuts.

(4) *Exterior unpainted metal surfaces.* Apply a coating of rust-preventive compound (heavy) to any mechanical or finished unpainted surfaces of the howitzer and carriage not already preserved.

g. Covers. Install all covers provided with the matériel and securely fasten.

h. Gun Book. (1) During storage or shipment, keep the gun book in a waterproof envelope securely fastened to the howitzer with nonhygroscopic adhesive tape.

(2) Under one of the wrappings of tape, insert one end of a small tab reading: "Gun book here."

i. Tires. (1) Remove all stones or other foreign objects from the tire treads.

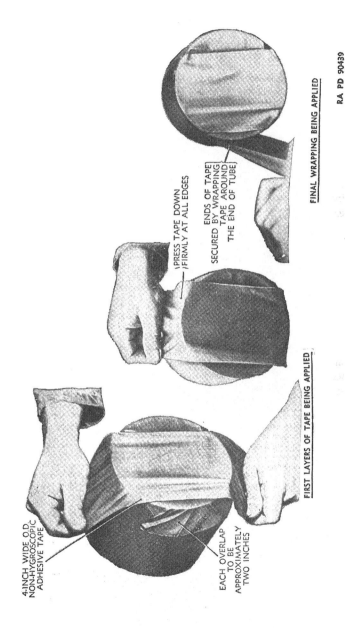

4-INCH WIDE O.D. NON-HYGROSCOPIC ADHESIVE TAPE

EACH OVERLAP TO BE APPROXIMATELY TWO INCHES

FIRST LAYERS OF TAPE BEING APPLIED

PRESS TAPE DOWN FIRMLY AT ALL EDGES

ENDS OF TAPE SECURED BY WRAPPING TAPE AROUND THE END OF TUBE

FINAL WRAPPING BEING APPLIED

RA PD 90439

Figure 122. Method of sealing muzzle.

(2) Keep rubber equipment free from oils, greases, and preservatives.

j. PREFERRED STORAGE. The preferred type of storage for this matériel is in closed dry warehouses or sheds. Where it is found necessary to store matériel in the open, cover with tarpaulins. Refer to SB 9–47.

k. STORAGE INSPECTIONS. (1) Before placing matériel in storage, make a systematic inspection as outlined in *c* above.

(2) Make a visual inspection periodically to determine general condition. If corrosion is found on any part, remove the rust spots, clean, and treat with the prescribed preservatives.

(3) Clean, inspect, and properly inflate all tires. Replace tires requiring repairing or retreading with serviceable tires. Matériel must not be stored on floors, cinders, or other surfaces, which are soaked with oil or grease. Wash off immediately any oil, grease, gasoline, or solvent which comes in contact with tires under any circumstances.

l. REMOVAL FROM LIMITED STORAGE. (1) If the matériel is not shipped or issued upon expiration of the limited storage period, further treat matériel for stand-by storage (matériel out of use for periods in excess of 90 days up to 3 years).

(2) Matériel being withdrawn from a storage status for shipment *will not* be deprocessed other than to insure that matériel is complete and serviceable. The removal of preservative will be the responsibility of the consignee.

(3) Deprocess matériel when it has been ascertained that it is to be placed into immediate service. Remove all rust-preventive compounds and thoroughly lubricate as prescribed in section XI. Thoroughly inspect matériel and service as prescribed in paragraphs 53 and 54.

3. Loading and Blocking Matériel on Railroad Car

a. LOADING RULES. (1) *Inspection.* Inspect railroad cars to see that they are suitable to carry loads to destinations. Floors must be sound and all loose nails or other projections not an integral part of the car must be removed.

(2) *Permanent ramps.* Use permanent ramps for loading the matériel when available. When such ramps are unavailable, use improvised ramps constructed of rail ties and/or other available lumber.

(3 *Handling.* (*a*) Cars loaded in accordance with specifications given herein must not be handled in hump switching.

(*b*) Do not cut off cars while in motion. Couple cars carefully and avoid unnecessary shocks.

(*c*) Place cars in yards or sidings so that they will be subjected to as little handling as possible. Designate separate track or tracks, when available, at terminals, classification or receiving yards for such cars. Cars must be coupled at all times during such holding and hand brakes set.

(4) *Clearance limits* of the railroads over which matériel is to be moved will govern the height and width of load. Army and railroad officials must check all clearances prior to each move.

(5) *Maximum load weights.* (*a*) In determining the maximum weight of load, the weights in Table VI will govern, except where load weight limit has been reduced by the car owner.

(*b*) For example, Table VI gives the capacity of the car as 100,000 pounds (column 1) and the total weight of car and load as 169,000 pounds (column 2). The permissible weight of load can be computed by subtracting the light weight of car, 37,000 pounds(stenciled on each side of car as "Lt. Wt.") from the total weight of car and load, 169,000 pounds (column 2). This gives a permissible weight of load of 132,000 pounds.

Table VI. Maximum Load Weights

(Col. 1) Marked capacity of car (lbs)	(Col. 2) Total weight of car and load* (lbs)
40,000	66,000
60,000	103,000
80,000	136,000
100,000	169,000
140,000	210,000
200,000	251,000

*The light weight of car, stenciled on each side of railroad car as "Lt. Wt.", must be subtracted from the figures in this column to obtain the permissible weight of load.

(6) *Brake wheel clearance* must be at least four inches below and six inches above, behind and to each side of the wheel (see A, fig. 123). Increase brake wheel clearance as much as is consistent with proper location of load.

(7) *Distribution of load.* Place load on the car so that there will not be more weight on one side of the car than on the other. One truck of the carrying car must not carry more than one-half of the load weight.

Note. Various types of matériel having common destinations may be loaded on the same car to require the use of a minimum number of cars.

(8) *Tire pressure.* Increase pressure to 10 pounds per square inch above normal for shipment by rail.

(9) *Type of car.* Use a flat, box, or drop-end gondola car.

(10) *Hand brakes.* Set brakes after loading the matériel.

(11) *Rotating parts of matériel.* It is imperative that the cradle traveling lock be securely fastened in order to prevent the tube from being elevated or traversed while the matériel is in transit.

b. BLOCKING MATÉRIEL ON RAILROAD CAR. (1) *General.* All blocking instructions specified herein are minimum and are in accordance with

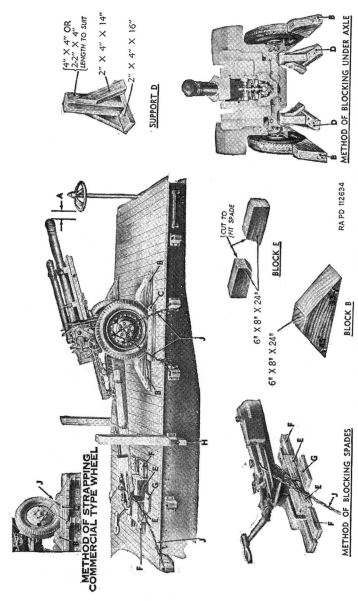

SUPPORT D

{4" X 4" OR
2-2" X 4"
(LENGTH TO SUIT

2" X 4" X 14"

2" X 4" X 16"

METHOD OF BLOCKING UNDER AXLE

BLOCK E

{CUT TO
FIT SPADE

6" X 8" X 24"

BLOCK B

6" X 8" X.24"

RA PD 112634

METHOD OF BLOCKING SPADES

METHOD OF STRAPPING
COMMERCIAL TYPE WHEEL

Figure 123. Method of securing 105-mm howitzer M2A1 and carriage M2A2 on railroad car.

Special Supplement No. 1 of Association of American Railroads "Loading of Commodities on Open Top Cars." Blocking material will be constructed of hardwood, fir, spruce, long leaf yellow pine, birch, or hemlock, straight-grained and free from strength impairing knots. Add additional blocking as required at the discretion of the officer-in-charge. All item reference letters given below refer to the details and locations in figure 123.

Note. Any loading methods or instructions developed by any source, which appear in conflict with this publication or existing loading rules of the carriers, will be submitted to the Office, Chief of Ordnance, Washington 25, D. C.

(2) *Brake wheel clearance.* Matériel will be placed in the traveling position on railroad car with the cradle traveling lock securely fastened. The gun tube must clear the brake wheel as prescribed in $a(6)$ above.

(3) *Blocks B* (pattern 3, four required). Locate the 45-degree portion of blocks against the front and rear of wheels. Nail heel of blocks to car floor with three fortypenny nails and toenail one side of blocks to car floor with two fortypenny nails.

(4) *Cleats C* (pattern 2, four required, 2 in. x 4 in. x 36 in. cleats). Position two cleats against the outside face of each tire. Nail lower cleat to car floor with three fortypenny nails and top cleat to lower cleat and car floor with three fortypenny nails.

(5) *Supports D* (pattern 7, two required). Place support under the axle near the inside face of each wheel. Cut support one-quarter inch longer than the distance between the axle and car floor, to partially relieve weight from the tires. Nail each support to the car floor with six fortypenny nails.

(6) *Blocks E* (pattern 6, four required, 6 in. x 8 in. x 24 in. blocks). Place blocks against the front and rear of each spade. Toenail blocks to car floor with five fortypenny nails. Blocks must be cut to fit contour of spades.

(7) *Cleats F* (pattern 1, eight required, 2 in. x 4 in. x 12 in. cleats). Place two cleats against each block E. Nail lower cleat to car floor with three fortypenny nails, and top cleat to lower cleat and car floor with three fortypenny nails.

(8) *Cleats G* (pattern 1, two required, 2 in. x 4 in. x 12 in. cleats). Place one cleat against each side of spade, and nail to car floor with three fortypenny nails.

(9) *Side stakes H* (pattern 8, two required, 4 in. x 5 in. x 48 in. stakes). Locate in stake pocket of car, one-third the distance from the end of trail to the center of the wheels. Cut stake to extend 4 inches below pocket. Drive one fortypenny nail into each stake directly below, and clinch over outside of pocket.

(10) *Strapping J.* (*a*) *Wheels — combat type.* (Four pieces are required.) Wheels are to be positioned with one hole at the top of the wheel.

    1. Pass one wire consisting of four strands of No. 8 gage, black annealed wire through two openings in wheel (top and adjacent hole) and through stake pocket. Bring ends of wire together and twist-taut with rod or bolt to remove slack.

    2. Pass other wire through top and adjacent hole (opposite step *1* above) and through stake pocket. (Wires must cross each other near center of wheel.) Bring ends of wire together and twist-taut with rod or bolt to remove slack.

(*b*) *Wheels—commercial type.* (Two pieces are required.) Wheels are to be positioned even at center of wheel, horizontally, as shown in inset, figure 123. Pass one wire consisting of four strands of No. 8 gage, black annealed wire through hole in back of wheel, and out of opposite hole, crossing wires at center and attaching to stake pockets. Bring ends of wire together and twist-taut with rod or bolt to remove slack.

(*c*) *Trails.* Pass one wire consisting of six strands of No. 8 gage, black annealed wire around end of trail and through stake pockets on both sides of car. Bring ends together and twist-taut with rod or bolt to remove slack.

4. Loading Matériel for Amphibious Movements

a. GENERAL. (1) These instructions prescribe procedures, methods, and practices to be followed when uncrated matériel is hoisted in or out of vessels, and describes the proper attachment points for slings to permit the matériel to be hoisted in its normal traveling position.

(2) Install covers supplied with matériel and securely fasten.

(3) If operations embrace deep water fording, prepare matériel in accordance with TM 9–2853.

(4) For methods in stevedoring, refer to TM 55–310.

b. SLING METHODS. (1) *Hoisting unboxed matériel in and out of vessels.* Due to varying conditions encountered in the field, use any of the following procedures where applicable (see fig. 124).

(*a*) *Method I* employs the following materials:

    1. Cable slings (2) (heavy enough to support matériel).
    2. Shackles (4) (placed between lifting cables and slings).
    3. Lifting cables (4).
    4. Lifting hook (1).
    5. Cargo runners (cable running through head and heel block and fastened to winch).
    6. Spreaders (2) (4 in. x 6 in., length as required). Use spreaders between cables to obtain a better balance and to provide clearance between slings and matériel preventing damage when hoisted. Spreaders consist of two pieces of oak or

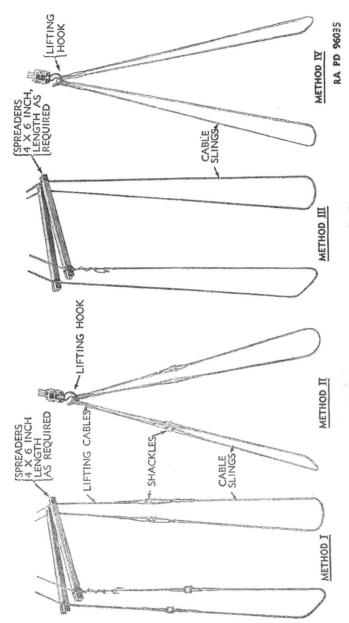

Figure 124. Sling methods used in hoisting artillery matériel.

RA PD 96035

other hardwood with open ends as shown in figure 124. Position on the cables leading from the lifting hook at a point which will provide clearance between slings and matériel. (The angle of the cable above the spreader must not exceed 45° from the vertical.) Slings are to be placed around matériel at the proper points of balance and attached to the shackles on the lifting cables.

(b) *Method II* requires special care to see that proper clearance for fire control brackets, gears, and operating surfaces is maintained when matériel is hoisted clear of ground. Method II employs the following materials:

1. Cable slings (2) (heavy enough to support matériel).
2. Shackles (4) (placed between lifting cables and slings).
3. Lifting cables (4).
4. Lifting hook (1).
5. Cargo runners (cable running through head and heel block and fastened to winch).

(c) *Method III* employs the following materials:
1. Cable slings (2) (heavy enough to support matériel).
2. Lifting hook (1).
3. Cargo runners (cable running through head and heel block and fastened to winch).
4. Spreaders (2) (4 in. x 6 in., length as required).

(d) *Method IV* requires special care to see that proper clearance for fire control brackets, gears, and operating surfaces is maintained when matériel is hoisted clear of ground. Method IV employs the following materials:

1. Cable slings (2) (heavy enough to support matériel).
2. Lifting hook (1).
3. Cargo runners (cable running through head and heel block and fastened to winch).

(2) *Cautions during hoisting of matériel.* (a) Before attempting to hoist matériel, examine hoisting cables to determine their condition. If strands of cable are broken at any point, substitute a new cable. Cables that are kinked and will not straighten out without damage *must not* be used.

(b) Under no circumstances must matériel be hoisted when it is found that all weight is balanced on one sling (other sling being loose). Lower matériel to the ground and place slings in the proper position.

(c) Do not place slings around howitzer tubes for hoisting purposes.

(d) Remove and securely stow all damageable instruments such as fire control equipment, sights, etc. *It is imperative that all stowage boxes, tires, or other loose equipment be securely strapped to matériel prior to movement.*

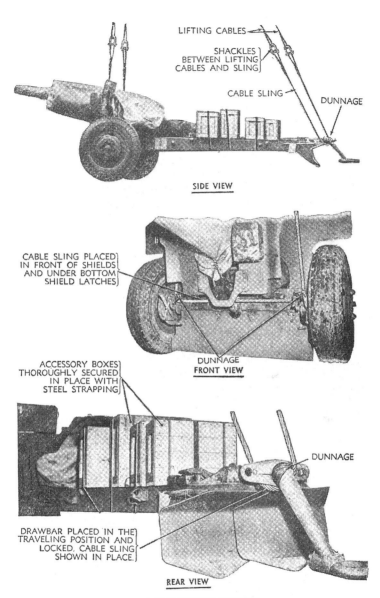

LIFTING CABLES

SHACKLES
BETWEEN LIFTING
CABLES AND SLING

CABLE SLING

DUNNAGE

SIDE VIEW

CABLE SLING PLACED
IN FRONT OF SHIELDS
AND UNDER BOTTOM
SHIELD LATCHES

DUNNAGE
FRONT VIEW

ACCESSORY BOXES
THOROUGHLY SECURED
IN PLACE WITH
STEEL STRAPPING

DUNNAGE

DRAWBAR PLACED IN THE
TRAVELING POSITION AND
LOCKED. CABLE SLING
SHOWN IN PLACE.

REAR VIEW

NOTE 1 — SUITABLE DUNNAGE WILL BE PLACED BETWEEN CABLE AND MATERIEL.
2 — ATTACH GUY LINES TO MATERIEL BEFORE LIFTING. GUIDE MATERIEL DURING
HOISTING TO PREVENT DAMAGE CAUSED BY STRIKING ANY NEARBY OBJECT
OR STRUCTURE.

RA PD 102947

Figure 125. Method of slinging 105-mm howitzer M2A1 and carriage M2A2.

191

(*e*) If the underside of matériel has sharp edges at the points where slings are placed, insert dunnage consisting of wood blocks, sacking, clean cloth, or similar material between the slings and matériel in order to prevent cable strands from cutting or slipping.

(*f*) Attach guy lines to matériel before lifting. Guide matériel during hoisting to prevent damage caused by striking any nearby object or structure.

(3) *Method of slinging 105-mm howitzer M2A1 and carriage M2A2* (figs. 124 and 125). (*a*) Place drawbar in the traveling position and lock.

(*b*) Position one cable sling in front of shields and under bottom shield latches.

(*c*) Position other sling between drawbar and end of trail.

(*d*) Locate dunnage as described in *b*(2)(*e*) above.

(*e*) Attach guy lines (*b*(2)(*f*) above), and hoist matériel slowly, observing proper balance (*b*(2)(*b*) above).

5. Disassembly and Crating Prior to Loading Matériel for Airplane or Truck Movement

a. PREPARATION FOR DISASSEMBLY. (1) *General.* Instructions contained in this paragraph cover the disassembly and crating prior to loading of the 105-mm howitzer M2A1 and carriage M2A2 in two C-47 transport planes or in a long wheelbase, 2½ ton, 6 x 6 truck (6 and 7, app. I). Employ the operations described herein *only* when it is impossible to transport matériel completely assembled; that is, jungle conditions where transportation is limited to men, animals, sleds, small boats, or aircraft.

Note. For information on disassembling and transporting the 105-mm howitzer M2 in the C-47 airplane, refer to TM 71-210.

(2) *Data.* (*a*) The matériel and equipment are disassembled and divided into 13 loads for transport. Four of these loads (2, 3, 4, and 9, see Table VII) have skid-type crates constructed to facilitate handling and loading operations. Loads 10 and 11 are placed in packing crates M32 and M33 (*1* below). The crates for these loads with their approximate weights are as follows:

	Approximate weight (lbs)
1. Crate, howitzer, M8 (fig. 126)	37
2. Crate, recoil and sleigh, M9 (fig. 127)	50
3. Crate, top carriage, M10 (fig. 131)	53
4. Crate, axle, M12 (fig. 132)	55
5. Crate, packing, M32 (figs. 133 and 134)	51
6. Crate, packing, M33 (figs. 135 and 136)	53

Note. Crates have been designed so that they may be constructed with the tools and lumber usually available to the battery. The wood used must be of firm texture such as pine, but soft enough to work with available tools. Hardwood will normally be beyond the capabilities of available tools and will unduly increase the weight. Due to manufacturing tolerances, crate dimensions may have to be modified slightly.

(*b*) Table VII shows the approximate weight and dimensions for each load.

Table VII. Thirteen Disassembled Loads for Transport

Loads No.	Item	Shipping weight (pounds)		Dimensions (inches)		
		Uncrated	Crated	Length	Width	Height
1	Shield	269	...	49	38	10
2	Barrel*	...	1005	105	12	13½
3	Recoil mechanism and sleigh*	...	511	72	12	22
4	Cradle, equilibrator and top carriage*	...	890	120	38	33
5	Right trail	380	...	156	19	19
6	Left trail	292	...	136	19	19
7	Right wheel	227	...	40	40	11
8	Left wheel	227	...	40	40	11
9	Axle assembly*	...	910	86	24	42
10	Miscellaneous parts*, composed of: Breechblock Firing lock Pintle pin Trail hinge pins Traversing mechanism Operating lever Elevating handwheel (right) Extractor Trigger shaft Operating lever pivot pin	...	225	29½	19½	10½
11	Miscellaneous parts*, composed of: Range quadrant Panoramic telescope mount	...	136	23	21½	14
12	Section chest	107	...	29	14½	12
13	Lifting bars and saw horses	131	...	...	...	...
	Total gross weight	1633	3677			

*Dimensions are for the crated matériel.

(3) *Cleaning.* Thoroughly clean the matériel (par. 2*d*, app. I) and lubricate (par. 2*e*, app. I), before disassembly is attempted. Select a site well protected from wind and loose dirt for cleaning operation.

(4) *Corrosion preventives.* (*a*) After disassembly of each unit, thoroughly clean all parts and dry prior to the application of corrosion preventives or lubricants.

(*b*) Coat all exposed metal surfaces with preservative lubricating oil (medium).

(*c*) Lubricate internal mechanisms in accordance with WD Lubrication Order Nos. 9–325 and 9–749, paragraph 42.

(*d*) To protect unpainted metal surfaces from direct contact with wood blocking or boxes, insert two layers of greaseproof wrapping and one layer of type C–1 waterproof barrier between the unpainted metal surfaces and the wood. To protect painted metal surfaces from contact, use any suitable cushioning material.

(*e*) Wrap finely machined parts in greaseproof wrapping and waterproof barrier material. Wiping cloths alone for wrapping and cushioning will not be used, as cloth will absorb any corrosion preventive which has been placed on a metal surface and will hold moisture against the surface, causing excessive rusting. Two layers of greaseproof wrapping and one layer of type C–1 waterproof barrier will be placed between cloth and metal to protect the surface of the parts. This is particularly necessary for the small parts which are wrapped in cloth and placed in packing crate M32 for miscellaneous parts.

(5) *Tools required for disassembly.* (*a*) *Standard tools*:

1 — Drift, brass, taper, $\frac{5}{16}$ inch pt., 4 inches long.
1 — Drift, brass, taper, $\frac{1}{2}$ inch pt., 6 inches long.
1 — Gun, lubricating, pressure, push-type, 5 oz.
1 — Hammer, hide-faced, 2 lb.
1 — Hammer, mach., ball-peen, 20 oz.
2 — Handles, socket wrench (wheel stud nuts).
1 — Jack, ratchet, 3-ton.
1 — Oiler, S., copper-plated.
1 — Pliers, side-cutting, flat-nose, 8 inch.
1 — Punch, drift, $\frac{1}{4}$ inch pt., 10 inches long.
1 — Punch, drive pin, std. S., $\frac{1}{8}$ inch pt.
1 — Screwdriver, comm., normal duty, 3 inch blade.
1 — Screwdriver, extra heavy duty, 5 inch blade (machs.).
1 — Wrench, engrs., dble-hd., alloy-S., $\frac{7}{16}$ inch x $\frac{1}{2}$ inch.
1 — Wrench, howitzer, locking ring.
1 — Wrench, screw, adj. knife handled, 12 inch.
1 — Wrench set, socket, hex. opening consisting of —
    1 — Bar, socket wrench ext. 5 inch.
    1 — Handle, ratchet, reversible.
    6 — Sockets, heavy duty, $\frac{7}{16}$ inch, $\frac{1}{2}$ inch, $\frac{9}{16}$ inch, $\frac{5}{8}$ inch, $\frac{3}{4}$ inch, $\frac{7}{8}$ inch.
2 — Wrenches, socket (wheel stud nuts).

(*b*) *Prime mover tools* ($2\frac{1}{2}$ ton, 6 x 6):

1 — Handle, lifting jack.
1 — Jack, hydraulic lifting 3-ton (if not available use jack, ratchet, 3-ton, from standard equipment).

1 — Pliers, combination, 6 inch.
1 — Wrench, adj. crescent, 12 inch.
1 — Wrench, screw, adj. 15 inch.
5 — Wrenches, engr., dble-hd., O.E. ⅝ x ²⁵⁄₃₂ inch, ¾ x ⅞ inch, ⅜ x ⁷⁄₁₆ inch, ½ x ¹⁹⁄₃₂ inch, ⁹⁄₁₆ x ¹¹⁄₁₆ inch.

(c) *Improvised tools*:

3 — Bars, wooden, lifting.
1 — Block, wooden, jack extension 8 to 10 inch section 1½ inch diameter.
1 — Guide, trail hinge pin (fig. 147).
2 — Saw horses, carpenter's wooden, large (approx., 30 x 30 inch).
2 — Saw horses, carpenter's wooden, small (approx., 20 x 20 inch) or appropriate wooden blocks.
1 — Wrench, pintle pin lower cover (fig. 129).

(6) *Order of disassembly.* The disassembly of the various components of the 105-mm howitzer M2A1 and carriage M2A2 is as follows:

(a) Shields (b below).

(b) Breech mechanism (c below).

(c) Howitzer barrel (c below).

(d) Recoil mechanism and sleigh (d below).

(e) Range quadrant (e below).

(f) Traversing mechanism (f below).

(g) Panoramic telescopic mount (g below).

(h) Pintle pin (h below).

(i) Cradle, equilibrator, and top carriage (i below).

(j) Trails (j below).

(k) Wheels with tires (k below).

(7) *Cautions for disassembly.* (a) Do not tap or drive steel against steel.

(b) Avoid force against fragile parts, such as range quadrant, sight mount, traversing and elevating handwheels.

(c) Reassemble nuts, bolts, pins, and keys (except pintle pin and trail hinge pins) to proper component promptly after the removal of each assembly. This prevents losing them and facilitates reassembly.

(d) Carefully inspect all removed assemblies for damage or unserviceable parts. Promptly report any such parts found in damaged condition to a responsible officer.

b. REMOVAL OF SHIELDS. (1) Remove nuts and bolts (with appropriate wrenches) which hold upper shields to brackets. The shields are removed in five segments with shield brackets left attached to the carriage and axle. One man must support each segment while another removes the nuts and bolts. This prevents the weight of the shield from resting against and damaging the threads on bolts.

(2) Remove cotter pins from shield apron hinge pins, driving pins out with hammer and brass drift.

(3) Remove the apron.

Note. For airplane loading, the stationary shields are bolted together face to face, and front shields are bolted to yoke which connects them when mounted on weapons.

c. REMOVAL AND CRATING OF BARREL. (1) *Removal.* Level the howitzer to facilitate easy removal and close the trails to allow a clear path on both sides for personnel walking to the rear with the barrel.

(*a*) Remove breech mechanism, thoroughly preserve and wrap (par. 2*f*(2) app. I) and place in packaging chest, M32 (*l* below, figs. 133 and 134).

(*b*) Loosen locking screw on the recoil mechanism bracket locking ring with a $\frac{7}{16}$-inch double head wrench (fig. 53), and remove the ring with a locking ring wrench (fig. 54).

(*c*) Place the howitzer crate, M8 (fig. 126), conveniently to the rear.

(*d*) Position four men on both sides of the howitzer with suitable lifting bars, and one man at the muzzle end with a wooden lifting bar. Start the howitzer moving to the rear by bunting the muzzle end with the wooden lifting bar. After howitzer starts moving, insert bar into the muzzle and guide it through the sleigh hoops. While four of the strongest men in the group support the breech end, remove howitzer and place it on the howitzer crate, M8, and thoroughly secure by strapping (fig. 126).

(*e*) Screw the recoil mechanism bracket locking ring back onto the tube and replace the lock screw.

(2) *Crate construction.* The crate and blocking for the barrel weighs approximately 37 pounds. The bill of material, Table VIII, shows the required number of pieces and dimensions of lumber. (See Note *a*(2)(*a*) above).

Table VIII. *Bill of Material for Howitzer Crate, M8*

Indicating letter	Quantity required	Name	Actual size—inches		
			Length	Width	Thickness
A	1	Bottom panel	72	12	2
B	3	Block	11½	4	2
C	2	Block (30 degree bevel on one end)	4	4	2
D	2	Block	15	2	2
E	1	Block	11½	4	1

d. REMOVAL AND CRATING OF RECOIL MECHANISM AND SLEIGH. (1) *Removal.* (*a*) Remove cotter pin and piston rod outer nut.

196

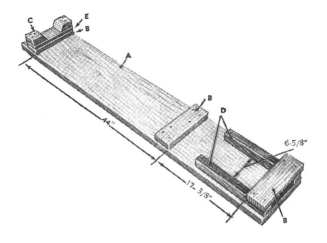

DETAILS OF SKID

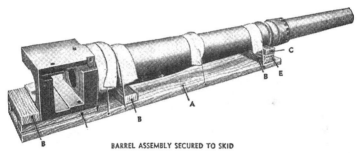

BARREL ASSEMBLY SECURED TO SKID

NOTE: FOR DIMENSIONS OF INDIVIDUAL PIECES REFER TO THE BILL OF MATERIAL.   RA PD 52492

Figure 126. Howitzer crate, M8.

(*b*) Push sleigh toward the rear.

(*c*) Remove the sleigh from the cradle with two lifting bars and place on the crate. Thoroughly secure by strapping (fig. 127).

(2) *Crate construction.* The crate and blocking for the recoil mechanism and sleigh weighs approximately 50 pounds. The bill of material, Table IX, shows the required number of pieces and dimensions of lumber. (See Note, *a*(2)(*a*) above).

e. REMOVAL OF RANGE QUADRANT. (See Cautions, *a*(7)(*b*) above) (1) Remove nuts and bolts which hold range quadrant M4 to the cradle bracket.

(2) Remove range quadrant and place in packing crate, M33 (see *l* below, and figs. 135 and 136).

197

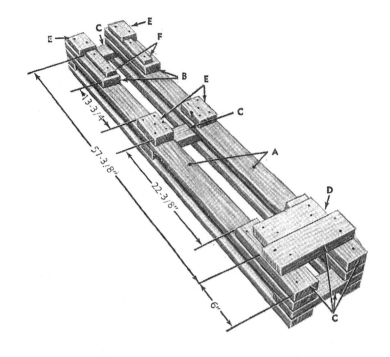

DETAILS OF SKID

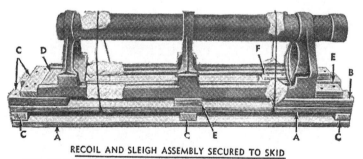

RECOIL AND SLEIGH ASSEMBLY SECURED TO SKID

NOTE: FOR DIMENSIONS OF INDIVIDUAL PIECES REFER TO THE BILL OF MATERIAL.

RA PD 82366

Figure 127. Recoil and sleigh crate, M9.

Table IX. *Bill of Material for Recoil Mechanism and Sleigh Crate, M9*

Indicating letter	Quantity required	Name	Actual size—inches		
			Length	Width	Thickness
A	4	Cleat (long)	72	4	2
B	2	Block	18	4	2
C	6	Block	12	4	2
D	1	Block	9½	4	2
E	4	Block	6	4	2
F	2	Block	9	2	1

f. REMOVAL OF TRAVERSING MECHANISM. (See Cautions, *a*(7)(*b*) above.) (1) Remove the cotter pin and nut (fig. 128) holding traversing swivel nut pivot to traversing handwheel screw bracket.

(2) Remove the screws and lock washers (fig. 128) holding traversing swivel nut bracket cap to traversing swivel nut bracket.

(3) Remove the traversing mechanism, replacing the caps, lock washers, and screws on the bracket, and place traversing mechanism in the packing crate, M32 (see *l* below, and figs. 133 and 134).

g. REMOVAL OF PANORAMIC TELESCOPE MOUNT. (See Cautions, *a*(7)(*b*) above.) (1) Remove the stud and two bolts holding the mount to the carriage.

(2) Loosen the lock screw in the castellated nut on the left trunnion, and remove nut and washer.

(3) Remove mount and place it in packing crate, M33 (see *l* below, and figs. 135 and 136). If the ring around the trunnion pin sticks, tap lightly with hammer and brass drift to loosen it.

h. REMOVAL OF PINTLE PIN. (1) Elevate the cradle.

Note. Counterbalance the front end of the cradle with three men in order to equalize the force of the equilibrator.

(2) Remove screws from the pintle pin lower cover with a $\frac{9}{16}$-inch socket wrench.

(3) Loosen and remove the cover with the pintle pin lower cover wrench (special tool, fig. 129).

(4) Remove the pintle pin from its housing by holding a wooden block against the bottom of the pintle pin, and with a lifting bar, lift the pintle pin upward as shown in figure 129.

(5) Remove the pintle pin key using brass drift and hammer.

(6) Place pintle pin and key in packing crate, M32 (*l*, and figs.133 and 134).

Note. At this point of disassembly, depress the cradle to a position approximately level.

TRAVERSING
SWIVEL NUT
PIVOT

TRAVERSING HANDWHEEL SCREW BRACKET
REAR

TRAVERSING SWIVEL
NUT BRACKET

FRONT

TRAVERSING SWIVEL
NUT BRACKET CAP

RA PD 82322

Figure 128. Removing traversing mechanism from carriage.

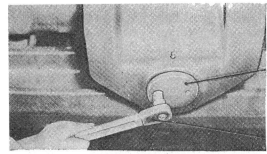

PINTLE PIN LOWER COVER

9/16" SOCKET WRENCH

REMOVAL OF SCREWS

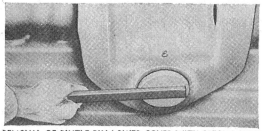

REMOVAL OF PINTLE PIN LOWER COVER WITH SPECIAL WRENCH

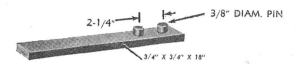

2-1/4" ⟶ ⟵ 3/8" DIAM. PIN

3/4" X 3/4" X 18"

SPECIAL TOOL (PINTLE PIN LOWER COVER WRENCH)

WOOD BLOCK

LIFTING BAR

REMOVAL OF PINTLE PIN FROM HOUSING   RA PD 82324

Figure 129.   Removal of pintle pin.

i. REMOVAL AND CRATING OF CRADLE, EQUILIBRATOR, AND TOP CARRIAGE. (1) *Removal.* (*a*) Spread the trails and position one small saw horse approximately two feet to the rear of the lower portion of the top carriage.

(*b*) Remove the cradle, equilibrator, and top carriage from the axle support by lifting with wooden bars. (This is accomplished by nine men, three men lifting at the front end with a fourth man pushing against the front of the cradle, and five men lifting at the rear.) Rest the lower portion of the top carriage on the small saw horse to the immediate rear.

(*c*) Support the cradle, equilibrator, and top carriage on the saw horse while the axle assembly, trails, and wheels are rolled forward about six to eight yards (fig. 130).

(*d*) Position a large saw horse underneath the rear of the cradle. Lift the front end of the cradle and remove the small saw horse, replacing with a large one. Then lower the front end onto the large saw horse placed immediately in front of the cradle lock strut latch.

(*e*) Clean and apply lubricating preservative oil (medium) to the miscellaneous parts, recoil mechanism, cradle, equilibrator, and top carriage assembly.

RA PD 27091

Figure 130. Supporting cradle, equilibrator, and top carriage while removing trails and axle.

(*f*) Before removing the elevating hand wheel, depress the elevating arcs with the aid of the auxiliary elevating hand wheel until the upper ends are flush with the cradle slides. Remove the elevating hand wheel.

(*g*) Lay the top carriage crate, M9, on the cradle and strap securely. Turn the assembly upside down (fig. 131).

Note. Often the cradle, equilibrator, and top carriage will be prepared for removal from the axle and support before the panoramic sight mount and range quadrant are dismounted. When this is done, the sight mount and range quadrant are removed after the cradle is placed upon the saw horses.

(2) *Crate construction.* The crate and blocking for the cradle, equilibrator, and top carriage weighs approximately 53 pounds. The bill of material, Table X, shows the required number of pieces and dimensions of lumber. (See Note, *a*(2)(*a*) above).

Table X. Bill of Material for Top Carriage Crate, M10

Indicating letter	Quantity required	Name	Actual size—inches		
			Length	Width	Thickness
A	1	Bottom panel	120	12	2
B	2	Block	11½	4	2
C	2	Block	8¼	4	2
D	2	Block	6½	4	2
F	2	Block	40	1½	⅝
E	10	Block (nailed together in pairs and placed under banding wires)	6	4	1

j. REMOVAL OF TRAILS. (1) Set the hand brakes and block wheels to prevent the axle and support from rolling backward when the trails are removed. Steady the assembly by holding the shield bracket as an added precaution.

Note. Before removal of trails, loosen wheel stud nuts on both wheels to facilitate removal of wheels and tires.

(2) Remove the cotter pin and castellated trail hinge pin nut.

(3) Remove the trail by tapping the trail hinge pin upward using a brass drift or short round section of wood and a hammer.

Note. If the pin has not been removed for sometime, it will probably be necessary to jack it out.

(4) Remove and replace trail hinge pin in packing crate, M32 (*l,* and figs. 133 and 134 below).

(5) Repeat same operations as described above for other trail.

k. REMOVAL AND CRATING OF AXLE ASSEMBLY. (1) *Removal of*

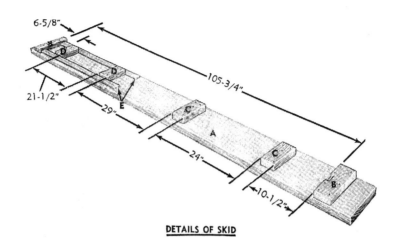

DETAILS OF SKID

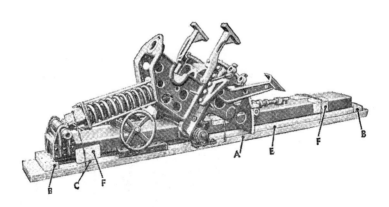

CRADLE, EQUILIBRATOR, AND TOP CARRIAGE SECURED TO SKID

NOTE: FOR DIMENSIONS OF INDIVIDUAL PIECES REFER TO THE BILL OF MATERIAL.

RA PD 82370

Figure 131. Top carriage crate, M10.

204

wheels and tires. (*a*) Raise each end of axle assembly so that tires clear the ground, and place saw horses underneath the axle.

(*b*) Remove wheel stud nuts that were loosened prior to removal of trails on both wheels and remove wheels.

(*c*) Replace stud nuts on hub assembly.

(*d*) Place axle assembly on crate and thoroughly strap in place (fig. 132).

(2) *Crate construction.* The crate and blocking for the axle assembly weighs approximately 55 pounds. The bill of material, Table XI, shows the required number of pieces and dimensions of lumber. (See Note, *a*(2)(*a*) above.)

Table XI. *Bill of Material for Axle Crate, M12*

Indicating letter	Quantity required	Name	Actual size—inches		
			Length	Width	Thickness
A	4	Skids (bevel)	72	4	2
B	10	Blocks (horizontal) (Notch out blocks as shown in figure 132)	24	4	2

i. CRATING OF MISCELLANEOUS PARTS. (1) *Parts for packing crates M32 and M33.* Below are listed the parts stowed in the packing crates, M32 and M33, and the bill of material for each crate.

(*a*) *Packing crate, M32* (figs. 133 and 134).

    1. Breechblock and firing lock (a leather thong or web belt around the breechblock provides a handle for insertion or removal from the box).

    2. Pintle pin.

    3. Trail hinge pins.

    4. Traversing mechanism.

    5. Operating lever.

    6. Elevating handwheel (right).

    7. Extractor.

    8. Trigger shaft. } Stowed underneath pintle and trail
    9. Operating lever pivot pin. } hinge pins.

(*b*) *Packing crate, M33* (figs. 135 and 136).

    1. Range quadrant.

    2. Panoramic telescope mount.

Note. If hinges and hasps are not available, the lids of the box for the miscellaneous parts may be secured by banding wire. Cleaning cloths may be placed in available space in crates.

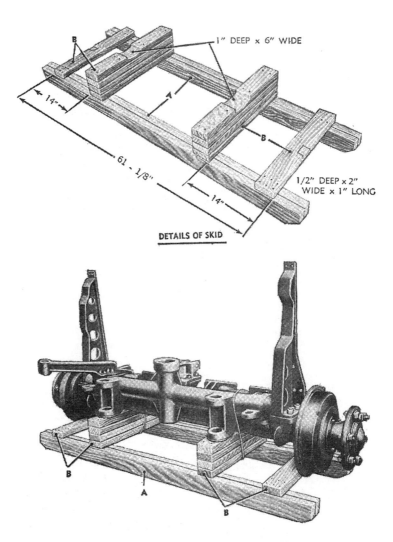

1" DEEP × 6" WIDE

B

14"

61 - 1/8"

B

14"

1/2" DEEP × 2"
WIDE × 1" LONG

DETAILS OF SKID

AXLE ASSEMBLY SECURED TO SKID

B

A

B

NOTE: FOR DIMENSIONS OF INDIVIDUAL PIECES REFER TO THE BILL OF MATERIAL.

RA PD 104969

Figure 132. Axle crate, M12.

(2) *Bill of material for packing crate, M32.* The crate together with blocking weighs approximately 51 pounds. The bill of material, Table XII, shows the required number of pieces and dimensions of lumber. (See Note, $a(2)(a)$ above.) Inside dimensions of crate with lid in place are 8¼ x 18 x 28 inches.

Table XII. Bill of Material for Packing Crate, M32

Indicating letter	Quantity required	Name	Actual size—inches		
			Length	Width	Thickness
		Crate[1]			
A	2	Side[2]	29½	7¼	¾
B	2	End[2]	18	7¼	¾
C	1	Bottom[2]	28	18	¾
D	2	Skid	19½	3¾	¾
E	1	Support	18	1⅝	3¾
F	1	Support	18	1⅝	3¾
G	1	Support	10¼	3¾	1¾
H	1	Support	10¼	3¾	1¾
J	1	Block	5½	3¾	1¾
K	1	Block	5½	4¾	¾
L	1	Support	13½	7¼	¾
M	1	Support	13½	7¼	¾
N	1	Block	2⅞	1½	¾
		Cover[1]			
O	1	Top[2]	28	18	¾
P	2	Top cleat	19½	3¾	¾
Q	2	Side[2]	28	1¾	¾
R	2	End[2]	19½	1¾	¾
S	4	Corner guide	2¹⁵⁄₁₆	1¾	¾
T	1	Pressure block	8½	3¾	1¾
U	1	Pressure block	4½	2¾	1⅝
V	1	Pressure block	3½	1⅝	1¾
W	2	Rope (handles)	24	¾ diam.	

[1]The crate must either be fastened together with long screws or reinforced on the corners with metal bands.

[2]Sides, end, top and bottom may be made of several pieces, but no piece shall be less than 2½ inches wide.

(3) *Bill of material for packing crate, M33.* The crate together with blocking weighs approximately 53 pounds. The bill of material, Table XIII, shows the required number of pieces and dimensions of lumber. (See Note, $a(2)(a)$ above.) Inside dimensions of crate with lid in place are 11 x 20 x 21½ inches.

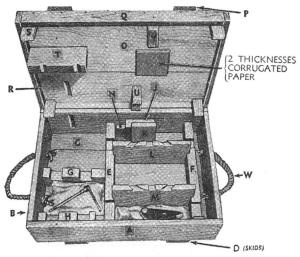

BLOCKING IN POSITION

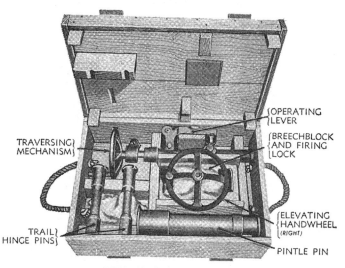

MISCELLANEOUS PARTS IN POSITION

NOTE: EXTRACTOR, TRIGGER SHAFT, AND OPERATING LEVER PIVOT PIN,
ARE STOWED UNDERNEATH PINTLE AND TRAIL HINGE PINS.

RA PD 82393

Figure 133. Packing crate, M32.

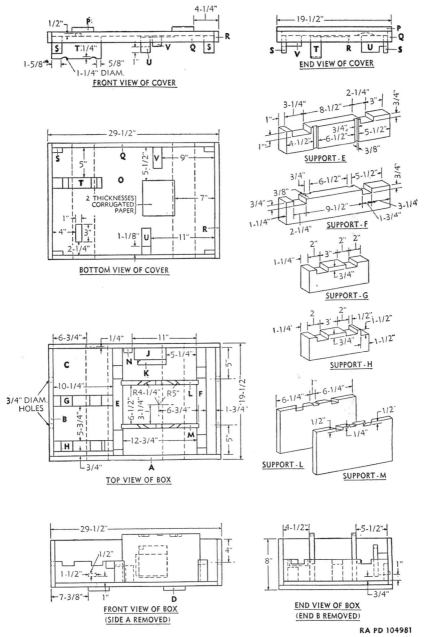

Figure 134. Details of packing crate, M32.

209

Table XIII. Bill of Material for Packing Crate, M33

Indicating letter	Quantity required	Name	Actual size—inches		
			Length	Width	Thickness
		Crate			
A	2	Side*	20	8	¾
B	2	End*	23	8	¾
C	1	Bottom*	21½	20	¾
D	2	Skid	21½	3¾	¾
E	1	Support	20	7¼	¾
F	1	Block	14	4	¾
G	1	Block	4	4	¾
H	1	Block	4	3½	¾
J	1	Block	4	2½	⅝
K	2	Block	4	3½	1¾
L	2	Block	5	3½	1⅝
M	1	Support	10	3½	1½
N	1	Support	9	3½	1½
O	1	Support	8	3⅝	1⅝
P	1	Block	4½	2⅝	1½
Q	1	Block	8	3¾	¾
R	1	Support	7	1½	1½
S	1	Block	8	4	⅝
		Cover			
T	1	Top*	23	21½	¾
U	2	Side*	21½	3¾	¾
V	2	End*	21½	3¾	¾
W	2	Top cleat	21½	3¾	¾
X	1	Pressure block	6	3⅝	1⅝
Y	1	Pressure block	3½	3⅝	1⅝
Z	1	Pressure block	5	2½	1⅝
AA	4	Corner guide	6	2	¾
BB	2	Rope (handles)	24	¾ diam.	

*Sides, ends, top and bottom may be made of several pieces, but no piece shall be less than 2½ inches wide.

6. Truck Loading (2½ Ton, 6 x 6)

a. Load the 105-mm howitzer M2A1 and carriage M2A2 with section chest, saw horses, and lifting bars in a 2½ ton, 6 x 6, long wheel base truck as shown in figure 137.

b. Prevent loads from shifting laterally by fitting 2 x 4 inch blocks between crate (fig. 137). Cut these blocks to fit and toenail to truck

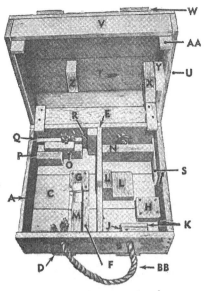

BLOCKING IN POSITION

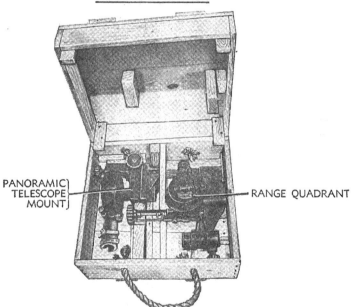

PANORAMIC
TELESCOPE
MOUNT

RANGE QUADRANT

MISCELLANEOUS PARTS IN POSITION

RA PD 104983

Figure 135. Packing crate, M33.

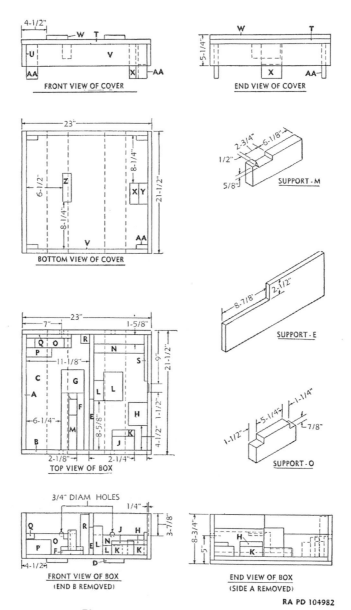

RA PD 104982

Figure 136. Details of packing crate, M33.

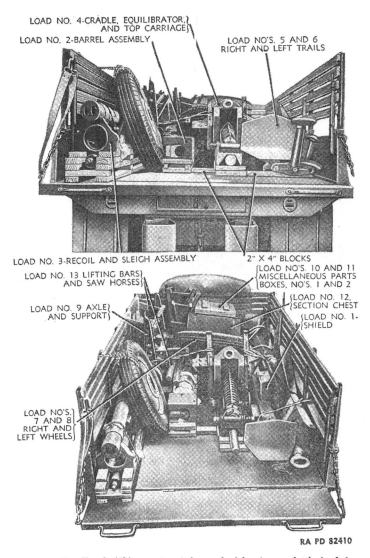

LOAD NO. 4-CRADLE, EQUILIBRATOR, AND TOP CARRIAGE

LOAD NO. 2-BARREL ASSEMBLY

LOAD NO'S. 5 AND 6 RIGHT AND LEFT TRAILS

LOAD NO. 3-RECOIL AND SLEIGH ASSEMBLY

LOAD NO. 13 LIFTING BARS AND SAW HORSES

LOAD NO. 9 AXLE AND SUPPORT

2" X 4" BLOCKS

LOAD NO'S. 10 AND 11 MISCELLANEOUS PARTS BOXES, NO'S. 1 AND 2

LOAD NO. 12, SECTION CHEST

LOAD NO. 1- SHIELD

LOAD NO'S. 7 AND 8 RIGHT AND LEFT WHEELS

RA PD 82410

Figure 137. Truck (2½ ton, 6 x 6, long wheel base) completely loaded.

flooring after howitzer is loaded. If the truck flooring is of steel construction, blocks shall be cut to form a tight wedge between matériel and/or crates in order to prevent shifting of loads.

c. Place packing crate, M32, across the tube and cradle, against the front end of the truck. Stack packing crate, M33, on top of it.

d. Tie trails to the side and forward corner of the truck, tie the shield brackets on the top carriage to the sides of the truck and tie one wheel (at rear) to the side.

Note. No. 8 gage soft iron wire is used for tying units in place.

e. Observe matériel during travel especially over rough terrain for any tendency of shifting to the rear as the tail gate cannot be closed.

7. Airplane Loading (Two C-47 Transport Airplanes)

a. GENERAL. For air transport, the use of two C-47 transport planes is recommended. The use of two airplanes permits carrying the howitzer and carriage, tools and equipment for reassembly, a crew of 10 men, and 32 rounds of ammunition.

b. LASHING. Suggested lashings are shown in figures 143 and 146 for both airplane loadings. The *baker bowline* knot is used to tighten lashings. Since the stability of the loads in the plane is entirely dependent on lashings, all personnel must be instructed in tying the baker bowline and method of tightening, if necessary, during air transit. All lashings are of $\frac{5}{8}$-inch rope and are 12 feet long.

Note. Two extra ropes in addition to those outlined in *c* and *d* below should be carried in each plane for emergency lashing during transit.

(1) *Two half-hitches.* The tie-down is prepared by tying rope to the tie-down ring by means of two half-hitches formed by bringing the free end of the rope through the tie-down ring. Pass the end around the rope and through its own loop to complete one half-hitch. Repeat the procedure in the same direction and then tighten complete knot (fig. 138).

(2) *Baker bowline.* The rope is passed through the tie-down ring and fastened to it with two half-hitches (fig. 139). Pass the running end of the rope "A" around the equipment to be lashed and back through tie-down ring (fig. 139). Form a bight or loop in the rope at "B" second step, (fig. 140). The rope is then grasped and folded as shown in third step (fig. 140). The fold is then passed up through the loop or bight, with the running end passing up through loop "C" (fig. 141). At this point pull on the running end "A" and secure equipment as tight as possible, then complete the baker bowline knob by fastening with a half-hitch around itself (fig. 141).

(3) *Slippery half-hitch.* The "slippery" half-hitch is a variation of the half-hitch, to facilitate untying the knot. It is simply a bight on a loop in the running end of the rope instead of a bight on the end itself (fig. 142).

Figure 138. Tying the double half-hitch.

c. AIRPLANE LOAD No. 1 (figs. 143 and 144). (1) The matériel to be loaded in the first C–47 transport plane with the number of ropes used for lashings are as follows:

(*a*) Barrel, recoil mechanism and sleigh (7 ropes).

(*b*) Cradle, equilibrator, and top carriage (4 ropes).

(*c*) Axle assembly (4 ropes).

(*d*) Ammunition (9 ropes).

(2) The order of loading with station number for center of gravity is given as follows:

	Station No.
(*a*) Recoil mechanism and sleigh	230
(*b*) Barrel	224
(*c*) Cradle, equilibrator, and top carriage	227
(*d*) Axle assembly	286
(*e*) Ammunition (ten rounds)	364
(*f*) Personnel (four men)	333

d. AIRPLANE LOAD No. 2 (figs. 145 and 146). (1) The matériel to be

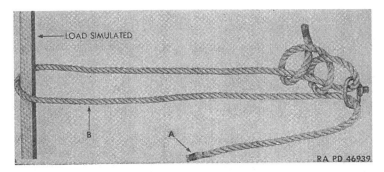

Figure 139. Starting the baker bowline knot.

215

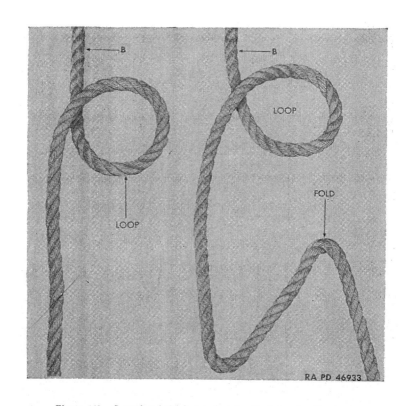

Figure 140. Second and third steps in tying the baker bowline knot.

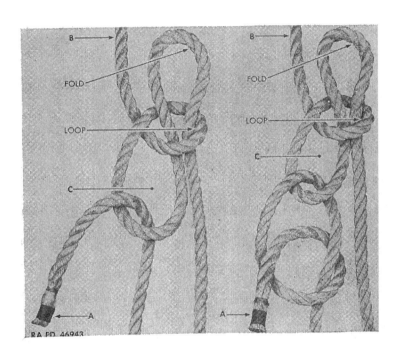

Figure 141. Completing the baker bowline knot.

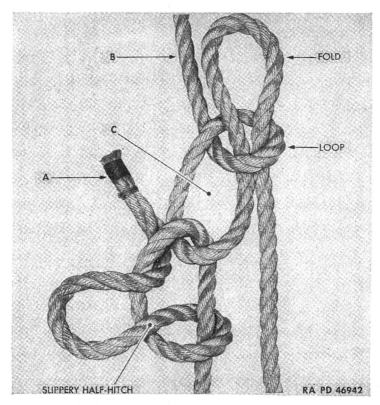

Figure 142. Tying the slippery half-hitch.

loaded in the second C–47 transport plane with the number of ropes used for lashings are as follows:

(*a*) Trails wheels packing crates, and saw horses (10 ropes).

(*b*) Shields section chest, and lifting bars (7 ropes).

(*c*) Ammunitio (10 ropes)

(2) The order of loading with station number for center of gravity is given as follows:

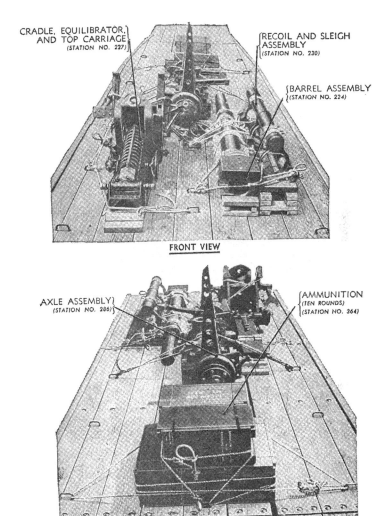

CRADLE, EQUILIBRATOR, AND TOP CARRIAGE (STATION NO. 227)

RECOIL AND SLEIGH ASSEMBLY (STATION NO. 230)

BARREL ASSEMBLY (STATION NO. 224)

FRONT VIEW

AXLE ASSEMBLY (STATION NO. 286)

AMMUNITION (TEN ROUNDS) (STATION NO. 364)

REAR VIEW

RA PD 88978

Figure 143. Load No. 1 positioned on C–47 transport plane mock-up front and rear.

219

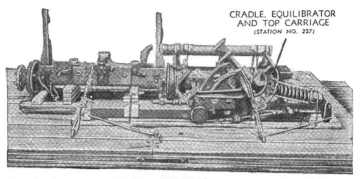

CRADLE, EQUILIBRATOR
AND TOP CARRIAGE
(STATION NO. 227)

RIGHT SIDE

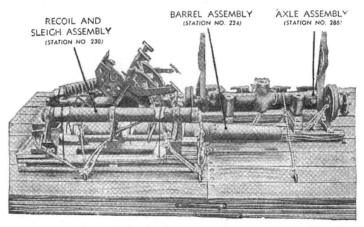

RECOIL AND
SLEIGH ASSEMBLY
(STATION NO 230)

BARREL ASSEMBLY
(STATION NO. 224)

AXLE ASSEMBLY
(STATION NO. 286)

LEFT SIDE   RA PD 88979

LOAD NO. 1

*Figure 144. Load No. 1 positioned on C–47 transport plane mock-up
left and right side.*

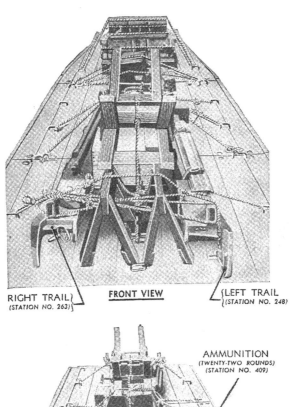

RIGHT TRAIL
(STATION NO. 263)

FRONT VIEW

LEFT TRAIL
(STATION NO. 248)

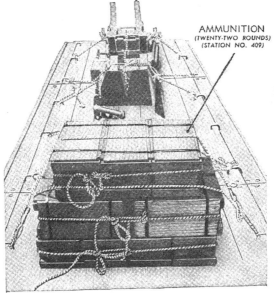

AMMUNITION
(TWENTY-TWO ROUNDS)
(STATION NO. 409)

REAR VIEW

RA PD 88980

*Figure 145. Load No. 2 positioned on C-47 transport plane mock-up
front and rear.*

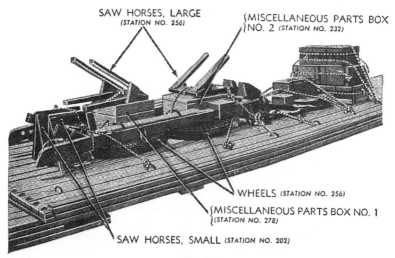

SAW HORSES, LARGE
(STATION NO. 256)

{MISCELLANEOUS PARTS BOX
{NO. 2 *(STATION NO. 232)*

WHEELS *(STATION NO. 256)*

{MISCELLANEOUS PARTS BOX NO. 1
{*(STATION NO. 278)*

SAW HORSES, SMALL *(STATION NO. 202)*

LEFT SIDE

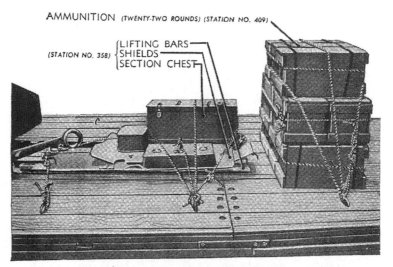

AMMUNITION *(TWENTY-TWO ROUNDS) (STATION NO. 409)*

(STATION NO. 358) {LIFTING BARS
{SHIELDS
{SECTION CHEST

VIEW SHOWING SHIELD, SECTION CHEST AND AMMUNITION

RA PD 88981

Figure 146. Load No. 2 positioned on C-47 transport plane mock-up left side.

8. Reassembly of Howitzer and Carriage

a. The matériel is reassembled in the reverse order of operations as described in paragraph 5b through k, appendix I. The trail hinge pin guide (fig. 147) is used in assembling the trails.

b. The holes in the top carriage and axle support through which the pintle pin is inserted are carefully alined by inspection before inserting the pin. It is usually necessary to pull down on the front end of the cradle to counteract the tendency of top carriage lower bearing to crowd forward, due to the rear heaviness which results when the fore part is lightened by removal of tube and sleigh. In combat operation, the pintle pin key must be removed (brass drift and hammer) when the howitzer is disassembled. As the absence of this key may slightly increase the wear of the pintle pin, it will be replaced at the earliest opportunity.

c. Before reassembly, all bearings, sliding surfaces, threads, exposed gears, etc., must be thoroughly cleaned, dried, lubricated, and inspected.

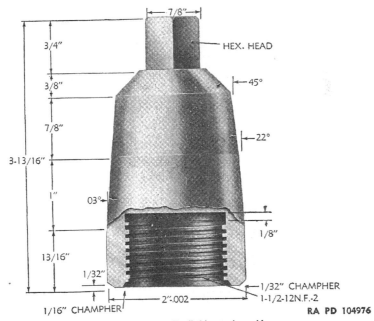

Figure 147. Trail hinge pin guide.

APPENDIX II

REFERENCES

1. Publication Indexes

The following publication indexes should be consulted frequently for latest changes or revisions of references given in this section and for new publications relating to matériel covered in this manual.

    a. List and Index of Department of the Army Publications FM 21–6

    b. List of War Department Films, Film Strips, and
        Recognition Film Slides .... FM 21–7

    c. Military Training Aids .... FM 21–8

    d. Ordnance Major Items and Combinations, and Pertinent
        Publications .... SB 9–1

    e. Ordnance Catalog Supply Index......WD Supply Catalog ORD 2

2. Standard Nomenclature Lists

    a. AMMUNITION

Ammunition, blank, for pack, light, and medium field,
    tank, and antitank artillery.. WD Supply Catalog ORD 11 SNL R–5

Ammunition, fixed and semi-fixed, including subcali-
    ber, for pack, light and medium field, aircraft, tank,
    and antitank artillery, including complete round
    data .... WD Supply Catalog ORD 11 SNL R–1

Ammunition instruction material for pack, light and
    medium field, aircraft, tank, and antitank artil-
    lery .... WD Supply Catalog ORD 11 SNL R–6

    b. HOWITZER MATÉRIEL.

Gun, 37-mm,
    subcaliber, M13 .. WD Supply Catalog ORD (*) SNL C–33, Sec. 16

Howitzer, 105-mm, M2A1; carriage, howitzer, 105-mm,
    M2A1 and M2A2; and mount, howitzer,105-mm,
    M4 and M4A1 .... WD Supply Catalog ORD (*) SNL C-21

    c. SIGHTING AND FIRE CONTROL EQUIPMENT.

Binoculars
    (various models) .. WD Supply Catalog ORD (*) SNL F–210, –238

(*) See WD Supply Catalog ORD 2, Index, for published pamphlets.

Chest, lighting equipment, M21; light,
aiming post, M14; light, instrument,
M19 .... WD Supply Catalog ORD (*) SNL F–205
Circle, aiming, M1..... WD Supply Catalog ORD (*) SNL F–160
Compass, M2 .... WD Supply Catalog ORD (*) SNL F–219
Finder, range, M7 w/F.A.
equipment ....WD Supply Catalog ORD (*) SNL F–254
Mount, telescope, M21A1, M23; Quadrant,
range, M4A1; Telescope, elbow, M16A1C,
M16A1D .... WD Supply Catalog ORD (*) SNL F–197
Mount, telescope, M42 .. WD Supply Catalog ORD (*) SNL F–256
Post, aiming, M1 ...... WD Supply Catalog ORD (*) SNL F–35
Quadrant, gunner's, M1.. WD Supply Catalog ORD (*) SNL F–140
Setter, fuze, M14....... WD Supply Catalog ORD (*) SNL F–1
Setter, fuze, M22....... WD Supply Catalog ORD (*) SNL F–293
Table, graphical firing,
M23 .... WD Supply Catalog ORD (*) SNL F–237
Telescope, battery commander's,
M65 .... WD Supply Catalog ORD (*) SNL F–259
Telescope, observation,
M48, M49 .... WD Supply Catalog ORD (*) SNL F–173
Telescope, panoramic,
M12A2 .... WD Supply Catalog ORD (*) SNL F–214
Thermometer, powder
temperature, M1 ..... WD Supply Catalog ORD 13 SNL M–5
Watch, pocket or wrist,
7-jewels or more ..... WD Supply Catalog ORD (*) SNL F–36

d. CLEANING AND PRESERVING.

Cleaning, preserving, and lubricating materials;
recoil fluids, special oils, and miscellaneous
related items .... WD Supply Catalog ORD 3 SNL K–1
Soldering, metallizing, brazing, and
welding materials, gases and
related items .... WD Supply Catalog ORD 3 SNL K–2

3. Explanatory Publications

a. AMMUNITION.
Ammunition, General .... TM 9–1900
Ammunition Inspection Guide .... TM 9–1904
Artillery Ammunition .... TM 9–1901
Ballistic Data, Performance of Ammunition............. TM 9–1907

(*) See WD Supply Catalog ORD 2, Index, for published pamphlets of the Ordnance Supply Catalog.

g. TRAINING FILMS AND FILM STRIPS.

The 105-mm Howitzer:

    Part I—Mechanical Functioning of the Howitzer...... TF 6–611
    Part II—Service of the Piece .... TF 6–612
Howitzer, 105-mm, M2A1 and Carriage, Howitzer,
    105-mm, M2:
    Part I—Basic Disassembly and Assembly .... FS 9–26
    Part II—Nomenclature, Disassembly and Assembly of
        Units, Inspection .... FS 9–27

4. Forms and Tables

Artillery Gun Book .... O.O. Form 5825
Firing Tables .... (See par. 113)
Unsatisfactory Equipment Report .... WD AGO Form 468

INDEX

229

230

231

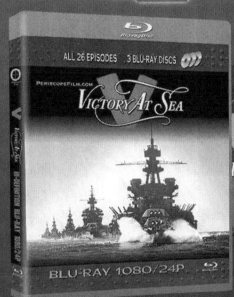

Made in the USA
Lexington, KY
26 November 2018